Intelligent Mobility in Smart Cities

Intelligent Mobility in Smart Cities

Editors

Leon Rothkrantz
Miroslav Svitek
Ondrej Pribyl

MDPI • Basel • Beijing • Wuhan • Barcelona • Belgrade • Manchester • Tokyo • Cluj • Tianjin

Editors
Leon Rothkrantz
Delft University of Technology
The Netherlands

Miroslav Svitek
Czech Technical University in Prague
Czech Republic

Ondrej Pribyl
Czech Technical University in Prague
Czech Republic

Editorial Office
MDPI
St. Alban-Anlage 66
4052 Basel, Switzerland

This is a reprint of articles from the Special Issue published online in the open access journal *Applied Sciences* (ISSN 2076-3417) (available at: https://www.mdpi.com/journal/applsci/special_issues/Mobility_Smart_Cities).

For citation purposes, cite each article independently as indicated on the article page online and as indicated below:

LastName, A.A.; LastName, B.B.; LastName, C.C. Article Title. *Journal Name* **Year**, *Volume Number*, Page Range.

ISBN 978-3-0365-4147-1 (Hbk)
ISBN 978-3-0365-4148-8 (PDF)

Cover image courtesy of Miroslav Svítek

Contents

About the Editors

Leon Rothkrantz

Leon Rothkrantz obtained his Master's degree in Mathematics from the University of Utrecht and his Master's degree in Psychology from the University of Leiden. He obtained a Ph.D. degree in Mathematics from the University of Amsterdam. In 2016, he obtained a Doctor honoris causa from the Technical University at Prague. From 1999 to 2008 he served as Associate Professor of Multimodal Communication at the Faculty Intelligent Systems from Delft University of Technology. From 2005 to 2008, he was the chairman of the Department Knowledge Base Systems from Delft University of Technology. In 2008, he was appointed as Professor of Intelligent Sensor-Systems at the Netherlands Defense Academy. From 2011 to 2015, he was Chairman of the Department Sensors Weapons and Command and Control. The topic of his research is Artificial Intelligence. He is (co-)author of more than 400 scientific papers.

Miroslav Svitek

Miroslav Svitek received the Ph.D. degree in radioelectronics at the Faculty of Electrical Engineering, Czech Technical University in Prague. Since 2005, he has been nominated as an extraordinary professor in applied informatics at the Faculty of Natural Sciences, University of Matej Bel in Banska Bystrica, Slovak Republic. Since 2008, he has been a full professor in engineering informatics at Faculty of Transportation Sciences, Czech Technical University in Prague. In 2010–2018. he was Dean of the Faculty of Transportation Sciences, Czech Technical University in Prague. Since 2018, he has been a visiting professor in smart cities at the University of Texas at El Paso, USA. The focus of his research includes complex system sciences and their practical applications to Intelligent Transport Systems, Smart Cities and Smart Regions. He is the author or co-author of more than 200 scientific papers and 10 books.

Ondrej Pribyl

Ondrej Pribyl, PhD is the Dean of the Faculty of Transportation Sciences at the Czech Technical University in Prague (CTU). He obtained his Master's degree at CTU, Faculty of Electrical Engineering in the area of technical cybernetics and artificial intelligence. He graduated at PennState University in the USA in the field of transportation planning. Since 2003, he has been working at the Faculty of Transportation Sciences (CTU), Department of Applied Mathematics. He became a full professor in the year 2017 and, in the year 2019, the head of the department. His professional interest is in diverse fields, including, among others, system theory, intelligent transportation systems, smart cities, mathematical modelling, data collection and analysis, or artificial intelligence. In the field of Smart Cities, he is working on simulation and modelling, travel behavior research (incl. the so-called activity-based approach), system theory and multi-agent systems.

Preface to "Intelligent Mobility in Smart Cities"

Smart Cities seek to optimize their systems by increasing integration through approaches such as increased interoperability, seamless system integration, and automation. Thus, they have the potential to deliver substantial efficiency gains and eliminate redundancy. To add to the complexity of the problem, the integration of systems for efficiency gains may compromise the resilience of an urban system. This all needs to be taken into consideration when thinking about Smart Cities.

The transportation field must also apply the principles and concepts mentioned above. This cannot be understood without considering its links and effects on other components of an urban system. New technologies allow for new means of travel to be built, and new business models allow for the existing ones to be utilized. This Special Issue puts together papers with different focuses, but all of them tackle the topic of smart mobility.

Leon Rothkrantz, Miroslav Svitek, Ondrej Pribyl
Editors

applied sciences

MDPI

Editorial

Intelligent Mobility in Smart Cities

Ondrej Pribyl [1,*], Miroslav Svitek [2] and Leon Rothkrantz [3]

1 Department of Applied Mathematics, Faculty of Transportation Sciences, Czech Technical University in Prague, Konviktská 20, 111000 Praha, Czech Republic
2 Department of Transport Telematics, Faculty of Transportation Sciences, Czech Technical University in Prague, Konviktská 20, 111000 Praha, Czech Republic; svitek@fd.cvut.cz
3 Department of Computer Science and Engineering, Delft University of Technology, Mekelweg 5, 2628 Delft, The Netherlands; l.j.m.rothkrantz@tudelft.nl
* Correspondence: author: pribylo@fd.cvut.cz

1. What Is a Smart City

In recent years, the term "Smart Cities" has become popular in Europe and abroad. There are various definitions of the term, but none of them are generally accepted. IBM was the first technologically leading company to discuss the Smart City Initiative in 2008. They introduced first concept of a Smart City:

> *"Cities are gaining greater control over their development, economically and politically. Cities are also being empowered technologically, as the core systems on which they are based become instrumented and interconnected, enabling new levels of intelligence. In parallel, cities face a range of challenges and threats to their sustainability across all their core systems that they need to address holistically. To seize opportunities and build sustainable prosperity, cities need to become smarter"* [1].

According to IEEE [2],

> *"A Smart City brings together technology, government and society to enable the following characteristics: a smart economy, smart mobility, a smart environment, smart people, smart living, and smart governance".*

The definition of European Commission, which promotes the general interests of the EU by proposing and enforcing regulatory compliance and implementing policies and the budget, is:

> *"A smart city is a place where traditional networks and services are made more efficient with the use of digital solutions for the benefit of its inhabitants and business. A smart city goes beyond the use of digital technologies for better resource use and less emissions. It means smarter urban transport networks, upgraded water supply and waste disposal facilities and more efficient ways to light and heat buildings. It also means a more interactive and responsive city administration, safer public spaces and meeting the needs of an ageing population"* [3].

The last example of definition of Smart City is from Department for Business Innovation & Skills:

> *"The concept is not static: there is no absolute definition of a smart city, no end point, but rather a process, or series of steps, by which cities become more „livable" and resilient and, hence, able to respond quicker to new challenges. Thus, a Smart City should enable every citizen to engage with all the services on offer, public as well as private, in a way best suited to his or her needs. It brings together hard infrastructure, social capital including local skills and community institutions, and (digital) technologies to fuel sustainable economic development and provide an attractive environment for all"* [4].

Citation: Pribyl, O.; Svitek, M.; Rothkrantz, L. Intelligent Mobility in Smart Cities. *Appl. Sci.* **2022**, *12*, 3440. https://doi.org/10.3390/app12073440

Received: 10 March 2022
Accepted: 16 March 2022
Published: 28 March 2022

When we look at the definitions, several key concepts can be extracted. Such selection of the major principles is summarized below:

1. Technology is a necessary tool, but it cannot be the purpose and aim of Smart Cities;
2. The focus is on providing services for citizens and making cities more livable (e.g., concepts such as 15-min cities);
3. Smart Cities seek to optimize their processes by increasing integration (i.e., synergic effect);
4. Building a Smart City is a process and cannot be simply bought;
5. Changing the way of thinking among not only citizens (e.g., ownership versus shared economy) but also decision makers (e.g., citizens' involvement in projects);
6. Smart city projects shall meet one or more of the following objectives
    a. Improving quality of life;
    b. Using resources more efficient;
    c. Improving city sustainability;
    d. Improving city resiliency and citizens' safety with focus on critical infrastructure.
7. Others

Smart Cities seek to optimize their systems by increasing integration through approaches such as increased interoperability, seamless system integration, and automation. Thus, they have the potential to deliver substantial efficiency gains and eliminate redundancy. To add to the complexity of the problem, it shall be stated that the integration of systems for efficiency gains may on the other hand compromise the resilience of an urban system. This all needs to be taken into consideration when thinking about Smart Cities.

2. How Is Mobility Changing within Smart Cities?

The transportation field must also apply the principles and concepts mentioned above. This cannot be understood without considering its links and effects on other components of an urban system. For example, the links between urbanism and transportation (the so called land use–transportation cycle [5]), urbanism and energy management, and many others shall be taken into consideration. New technologies allow new means of travel to be built, and new business models allow existing ones to be utilized. There is a whole paradigm shift in this context from focusing on mobility (i.e., the ability to travel) and accessibility (i.e., the ability to participate in activities).

While we cannot discuss all of such trends and principles, an overview prepared by the author is provided in the Figure 1.

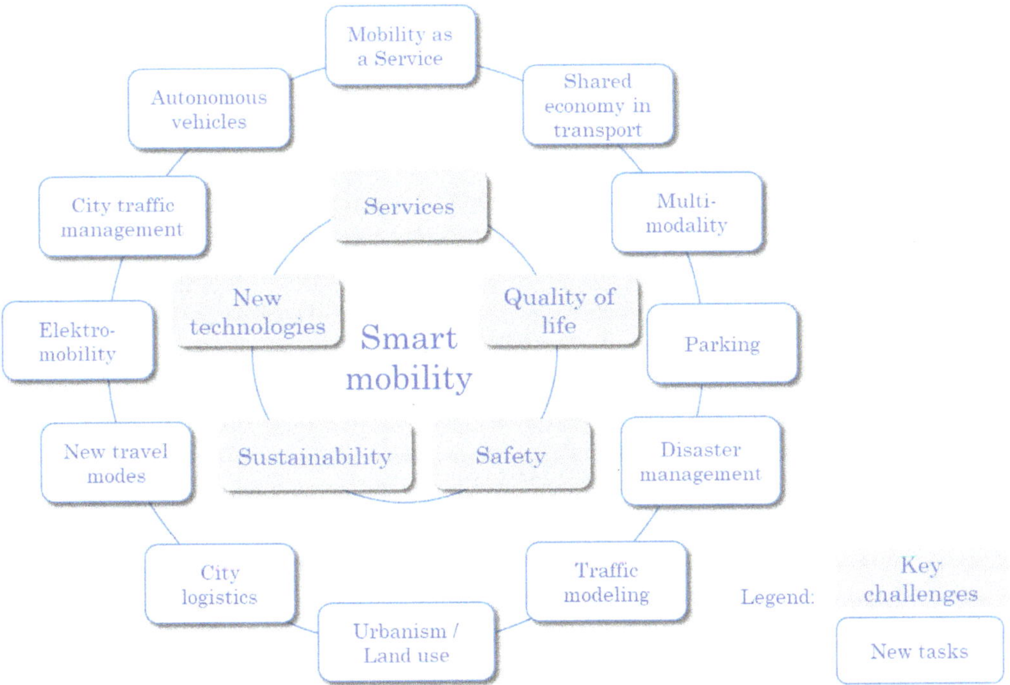

Figure 1. Overview of new trends and challenges within smart mobility (own source).

3. The Approach of This Special Issue

This Special Issue puts together papers with different focuses, but all of them tackle the topic of smart mobility as defined and discussed above.

The topic of integration of mobility into the wider city perspective is addressed within the paper "Interdisciplinary Urban Tunnel Control within Smart Cities" [6]. The paper has two major contributions. First, it provides a systematic view of an urban road tunnel with a focus on the interfaces between the tunnel and the rest of the city and the way they are managed. A tool to consider a sustainable development of a tunnel (i.e., not only traffic flow parameters such as average speed but also environmental and societal characteristics) is provided. The second contribution is that it provides a new urban road tunnel control approach that follows the original methodology and systemic view described in the paper. If the tunnel is controlled autonomously, which corresponds to the current state of the art in many cities, the algorithm decides to close it based on only local parameters. However, the proposed new algorithm takes into consideration not only the traffic situation in the tunnel (as expressed by the parameter traffic density) but also the actual traffic situation within the city (as expressed by its level of service (LOS)). This allows more environmentally, socially, and economically sustainable management of urban road tunnels.

The topic of city logistics using new and environmentally friendly travel modes is addressed in the paper "A Geometrical Structure-Based New Approach for City Logistics System Planning with Cargo Bikes and Its Application for the Shopping Malls of Budapest" [7]. In this paper, the authors provide a new geometrical-model-based way of thinking in organizing city logistics by the use of cargo bikes. Although cargo bikes are popular in many countries particularly in Asia, their usage in European cities is not common. A new approach to managing city logistics is by clearly aiming to improve city sustainability. Within this paper, a model is provided to plan the location of cargo bikes within the capital of Hungary, i.e., Budapest.

Another contribution to the topic of city logistics is in the paper "A Multi-Criteria Decision-Making Approach for Ideal Business Location Identification" [8]. It searches for optimal placement of business within a city. This has a positive effect on the accessibility of activities, as discussed above, and can contribute, for example, to decreasing the number of miles traveled by vehicles, which is an important performance indicator. The paper focuses on the decision-making process and presents a hybrid of two techniques to solve the MCDM problem. The proposed approach is based on the AHP (Analytic Hierarchy Process) and TOPSIS (The Technique for Order of Preference by Similarity to Ideal Solution) approaches. The hybrid approach reduces the computational complexity and requires less manual effort, thus improving the efficiency and accuracy of the proposed approach. Given a set of candidate locations for a new business, the proposed approach ranks the candidates. Thus, the candidate locations with higher ranks are identified as suitable or ideal. Such a decision-making model has the potential to also be applied to other aspects of Smart Cities.

The focus on citizens through services is also one of the key objectives mentioned above. This is true especially when looking at seamless integration of various travel modes. The paper "Exploring Hybrid-Multimodal Routing to Improve User Experience in Urban Trips" [9] introduces a new hybrid-multimodal routing algorithm that evaluates different routes that combine different transportation modes. Hybrid-multimodal routes are route options that might consist of more than one transportation mode. The motivation to use different transportation modes is to avoid unpleasant trip segments (e.g., traffic jams, long walks) by switching to another mode. Another contribution of the paper is in the approach of how to deal with geo-located and anonymized data.

An improvement in the efficiency of transportation services is also enabled through the connectivity of various actors within the field. This belongs to the field of Cooperative Intelligent Transport Systems (C-ITS). The term refers to transport systems, where the cooperation between two or more ITS sub-systems (personal, vehicle, roadside, and central) enables and provides an ITS service that offers better quality and an enhanced service level compared to the same ITS service provided by only one of the ITS sub-systems. Today, pilot implementations of cooperative systems in European countries are being carried out. However, before they are put into full operation, they need to be tested, evaluated, and assessed. The paper "Methodology of Functional and Technical Evaluation of Cooperative Intelligent Transport Systems and Its Practical Application" [10] focuses on the latter two points, i.e., evaluation and assessment of the cooperative systems. For this purpose, a methodology was created that describes the procedure chosen in the evaluation and assessment of cooperative systems in the Czech Republic and a demonstration of its use by way of example. The methodology is focused on three main areas, which in this case are functional evaluation, user acceptance, and impact assessment.

The last paper of this Special Issue, "Comprehensive Analysis of Housing Estate Infrastructure in Relation to the Passability of Firefighting Equipment" [11], focuses on urban emergency management (i.e., resiliency). It provides the assessment and evaluation of the passability in densely populated parts of cities with multi-story housing estates in terms of the operation of the integrated rescue system (IRS) in the Czech Republic. The aim of the research is to minimize the arrival times in order to conduct the intervention as efficiently as possible. The presented problem is caused by the unsystematic development of housing estates and the emergence of secondary problems in the form of the inability of larger IRS vehicles to reach the place of intervention. The vision presented in this document presents a systematic approach to improve the serviceability of individual blocks of flats. The main aim is to ensure passability, even for the largest types of equipment such as fire engine ladders.

While this Special Issue does not solve the problem of mobility within Smart Cities, we believe the papers make an interesting contribution to the field of transportation.

Institutional Review Board Statement: Not applicable.

Informed Consent Statement: Not applicable.

Data Availability Statement: Not applicable.

Acknowledgments: Thanks are due to all the authors and peer reviewers for their valuable contributions to this Special Issue. The paper is a result of the research done in the project "Smart City–Smart Region—Smart Community"—project (No. CZ.02.1.01/0.0/0.0/17 048/0007435) financed by Czech Operational Programme "Research, Development and Education" for the implementation of the European Social Fund (ESF) and the European Regional Development Fund (ERDF).

Conflicts of Interest: The authors declare no conflict of interest.

References

1. IBM. A Vision of Smarter Cities. Available online: http://www-01.ibm.com/common/ssi/cgi-bin/ssialias?infotype=PM&subtype=XB&appname=GBSE_GB_TI_USEN&htmlfid=GBE03227USEN&attachment=GBE03227USEN.PDF/ (accessed on 8 March 2022).
2. IEEE Smart Cities. About Smart Cities. Available online: https://www.ieee-pes.org/pes-communities/ieee-smart-cities (accessed on 8 March 2022).
3. European Commission: Digital Agenda for Europe. Smart Cities: A Europe 2020 Initiative. Available online: https://ec.europa.eu/info/eu-regional-and-urban-development/topics/cities-and-urban-development/city-initiatives/smart-cities_en (accessed on 8 March 2022).
4. Department for Business Innovation & Skills. Smart Cities: Background Paper. Available online: https://assets.publishing.service.gov.uk/government/uploads/system/uploads/attachment_data/file/246019/bis-13-1209-smart-cities-background-paper-digital.pdf (accessed on 8 March 2022).
5. Rodrigue, J.-P. *The Geography of Transport Systems*, 5th ed.; Routledge: New York, NY, USA, 2020; p. 456, ISBN 978-0-367-36463-2.
6. Přibyl, O.; Přibyl, P.; Svítek, M. Interdisciplinary urban tunnel control within smart cities. *Appl. Sci.* **2021**, *11*, 10950. [CrossRef]
7. Sárdi, D.L.; Bóna, K. A geometrical structure-based new approach for city logistics system planning with cargo bikes and its application for the shopping malls of Budapest. *Appl. Sci.* **2021**, *11*, 3300. [CrossRef]
8. Shaikh, S.A.; Memon, M.; Kim, K.-S. A multi-criteria decision-making approach for ideal business location identification. *Appl. Sci.* **2021**, *11*, 4983. [CrossRef]
9. Rodrigues, D.O.; Maia, G.; Braun, T.; Loureiro, A.A.F.; Peixoto, M.L.M.; Villas, L.A. Exploring hybrid-multimodal routing to improve user experience in urban trips. *Appl. Sci.* **2021**, *11*, 4523. [CrossRef]
10. Lokaj, Z.; Šrotýř, M.; Vaniš, M.; Mlada, M. Methodology of functional and technical evaluation of cooperative intelligent transport systems and its practical application. *Appl. Sci.* **2021**, *11*, 9700. [CrossRef]
11. Vrtal, P.; Kohout, T.; Nováček, J.; Svatý, Z. Comprehensive analysis of housing estate infrastructure in relation to the passability of firefighting equipment. *Appl. Sci.* **2021**, *11*, 9587. [CrossRef]

applied sciences

MDPI

Article

Interdisciplinary Urban Tunnel Control within Smart Cities

Ondřej Přibyl *, Pavel Přibyl and Miroslav Svítek

Faculty of Transportation Sciences, Czech Technical University in Prague, Konviktská 20, 110 00 Praha 1, Czech Republic; pribyl@fd.cvut.cz (P.P.); svitek@fd.cvut.cz (M.S.)
* Correspondence: pribylo@fd.cvut.cz

Abstract: Nowadays, urban road tunnels are considered to be independent entities within a city. Their interactions with the rest of the city and vice versa are usually not considered and, if they are, are only considered in a limited way (for example, through the nearest traffic controller). Typically, only the traffic parameters and not the environmental impacts are considered. This paper has two major objectives. First, we provide a systemic view on a road urban tunnel. The major focus is on the interfaces between the tunnel and the rest of the city and the way they will be managed. We are providing a tool to take into consideration a sustainable development of a tunnel (i.e., not only traffic flow parameters such as average speed, but also environmental and societal characteristics). This model expresses the actual traffic situation in a monetary form (i.e., cost of congestions). The second objective is to provide a new road urban tunnel control approach that follows the original methodology and systemic view described in the paper. If the tunnel is controlled autonomously, which corresponds to the current state-of-the-art in many cities, the algorithm decides to close it based on only local parameters. However, the proposed new algorithm takes into consideration not only the traffic situation in the tunnel (expressed by the parameter traffic density), but also the actual traffic situation within the city (expressed by its level of service (LOS)). This allows more environmentally, socially and economically sustainable oriented road urban tunnel management. The described algorithm is demonstrated on a specific example of the tunnel complex Blanka in Prague.

Keywords: traffic control; urban tunnel; smart cities; integration

Citation: Přibyl, O.; Přibyl, P.; Svítek, M. Interdisciplinary Urban Tunnel Control within Smart Cities. *Appl. Sci.* **2021**, *11*, 10950. https://doi.org/10.3390/app112210950

Academic Editor: Luís Picado Santos

Received: 29 October 2021
Accepted: 17 November 2021
Published: 19 November 2021

1. Introduction

A road tunnel is often denoted a complex system, since the documentation contains hundreds of pages and consists of several large sub-systems. To describe it, it is necessary to get together experts with various backgrounds including civil engineers, electrical engineers, and also environmentalists or traffic engineers. At the same time, the design of an urban road tunnel is based not only on the national or international standards and regulations, but also on mathematical and physical models and designers' experience. There is not a mandatory prescription on how to further split the tunnel into subsystems or how many and which processes must be implemented. For example, the tunnel's safety system involves at least the followings subsystems: SOS boxes, identification of dangerous goods through ADR (hazard identification number) code recognition, powered PA speakers, video-detection of pedestrians, accident or smoke detection, section speed control, and others. Each of them is composed by diverse facilities.

All of the above mentioned indicates that a tunnel is a dedicated and separated system. In this paper we highlight the importance of recognizing an urban tunnel as part of a city. There are many processes that are interconnected. As will be demonstrated, a tunnel control strategy influences traffic in the rest of the city.

The main problems of integration an urban tunnel into a (smart) city is the large number of processes, their specific features, and the complicated relationship between them which all lead to the increase complexity and create a high dimensionality problem making it difficult to share.

The main contribution of this paper is in two aspects. First, the role of an urban road tunnel within a city is discussed. The topic of smart cities looks at quality of life and sustainability usually through better integration of their components [1]. A tunnel is, however, nowadays considered an independent entity. For this reason, we discuss the different dimensions of the interface between a tunnel and a city. We provide a target function as well as a mechanism to express the traffic situation in monetary terms. This allows better integration.

The second contribution of this paper is a new tunnel control algorithm based on the systemic approach described above. This algorithm takes into consideration the future state of traffic flow in the tunnel. It is based on the measured parameter traffic density in the tunnel but also as on the level of service (LOS) of traffic in the rest of the city. Thus, the control is adopted based not only on the tunnel performance itself but also on the rest of the city. A parameter cost of congestions is explained and shall be used directly to evaluate the control strategies. This parameter also demonstrates the monetary influence of current tunnel control strategies (namely closing the tunnel for road traffic) on the city.

This paper has the following structure. First, we address an urban road tunnel from the system theory perspective. The target function of any control or management strategy is discussed. Also, the topic of complexity and interface to the adjacent (smart) city components and processes is addressed. The influence of tunnel control on the adjacent traffic is supported by results from a project GLOMODO (global model of traffic) on a case study from Prague. The theory of expressing the deterioration of traffic performance indicators is expressed through monetary terms (i.e., cost of congestions). In Section 3, we discuss a new urban tunnel control algorithm. This algorithm is based on the traffic flow theory and aims on keeping the flow stable (fluid) while taking into consideration the outside traffic of the tunnel. The impact is demonstrated on a case study showing the effect of closing a tunnel complex Blanka in Prague. This is followed by a discussion and conclusions.

2. Systemic Approach to Urban Road Tunnel

2.1. Tunnel and Its Neighborhood

In the literature, a tunnel is often described as a complex system. Even the World Road Association (PIARC) in its PIARC Road Tunnels Manual [2] from 2019 dedicated an entire chapter to the topics of complexity. It is called "TUNNEL: A COMPLEX SYSTEM". The authors claim that *"A tunnel constitutes a complex system which is the result of the interaction of very many parameters. These parameters can be gathered by subsets."* This is certainly correct. A tunnel consists of many different subsystems and components that interact with each other. Can we, however, claim that it truly is a complex system?

Let us look at classification based on complexity theory. Rzevsky and Skobelev [3] provide seven criteria a system shall fulfil in order to be considered a complex system. In the following paragraphs, we will provide these criteria together with a discussion addressing the road tunnels in particular:

1. Connectivity—A complex system consists of a large number of diverse components, known as Agents, which are richly interconnected.

Tunnels consist of different components and subsystems. We can perceive all of the different sensors and actuators as agents. It cannot be, however, claimed that they are "richly interconnected." They have interfaces and interact with each other. These interfaces and the reaction to changes of the shared variables are however clearly predefined.

2. Autonomy—Agents are not centrally controlled; they have a degree of autonomy but their behavior is always subject to certain laws, rules or norms.

This criterion is not fulfilled. All subsystems have a predefined behavior that is centrally controlled. There is a typically a central control system (SCADA—supervisory control and data acquisition) which access all the inputs from all subsystems, provides control decision and affects the actuators/outputs.

3. Emergence—*Global behavior of a complex system emerges from the interaction of agents and, in turn, constrains agent behavior. Emergent behavior is unpredictable but not random.*

The emergence in behavior of a complex system is its crucial characteristic. Based on the behavior of partial agents, one cannot determine the overall behavior of the system. This is clearly not true for existing road tunnels. Here, the administration expects clear and predictable behavior. Without that, a road tunnel would not be put into operation.

4. Nonequilibrium—*Complex systems generate unpredictable disruptive events.*

Similarly to the previous criterion, the purpose of tunnel control system is to avoid disruptive events. Even in case of a traffic accident, there are predefined rules that are addressing such situation. The objective is to minimize any disruptiveness.

5. Nonlinearity—*Relations between agents are nonlinear.*

This criterion is fulfilled as for example interaction between traffic parameters and environmental characteristics are nonlinear. This is true for most of the partial subsystems.

6. Self-Organization—*Complex systems self-organize, i.e., autonomously change their behavior or modify their structure, to eliminate or reduce impact of disruptive events (adaptability) or to repel attacks (resilience).*

Tunnel systems do not self-organize. In order to ensure predictability (again, this is a characteristic expected by the road administrators), the behavior even in case of for example traffic accident is predefined.

7. Co-Evolution—*If we define the system environment as the set of all systems with which the system interacts, then we can postulate that complex systems are open, they adapt to their environments and, in turn, change their environments.*

As by self-organization, the existing control systems do not adapt to their environment. They can (and should) be adapted to the changes in environment by the system designers, but it is not a default property of the system.

The arguments above showed that we cannot address road tunnel system as a "complex system" according to the system theory. But is it possible? Nobody can claim that a road tunnel is a simple system. It covers other aspects, such as environment, traffic management, energy management, or even emergency management. So how come it is not complex? It is important to understand that the opposite "complex" isn't "simple", but "independent"; the opposite of "simple" is "complicated" [4]. And road tunnel system is certainly a complicated system, with many subsystems.

Another aspect that contributed to the confusion in terms is the linguistics. We talk often about a "tunnel complex" or as in case of the above-mentioned Road Tunnels Manual [2]: "Complex underground road networks". That is correct. It is not just a hole in a hill. It covers all the different subsystems discussed above; often there are successive connected tunnels (see examples in Prague, The Hague or Oslo; the tunnels are often multimodal (e.g., shared usage between buses, pedestrians, bicycles and trams); tunnels can have a dual function (as transit and access to underground car parks—e.g., Annecy, Brussels and Tromsø); and others). Without a doubt, we are talking about "tunnel complex". Again, it is not the same as a tunnel being a "complex system".

In this paper, however, we will not assume that the urban tunnel is the element itself, but on the contrary part of the smart city concept. In this connection, it becomes a complex system with its neighborhood according to the described principles, since the tunnel must communicate with other subsystems of the smart city. Within this communication, the tunnel becomes an autonomous agent which negotiates with other systems (also represented by software agents) the best possible method for its management.

2.2. Sustainable Target Functions of an Urban Tunnel System

Tunnels are a specific part of the road network. Often, they are considered a closed system. We will demonstrate the need for their better integration not only into the road traffic network, but also to other fields such as energy management and environmental impacts.

It is more expensive to reach the same safety level in tunnels than on open roads. In case of an accident, the emergency response is far more difficult and human lives are in danger more often [5].

The primarily objective of road tunnels is to enhance the urban road network and thus help to realize transport performance between two points (origin-destination) expressed in vehicle kilometers travelled (VKT).

The transport performance however cannot be the only measure of quality. The tunnel (actually similarly to the rest of the road network), must follow the principles of sustainable development [6] as depicted in Figure 1.

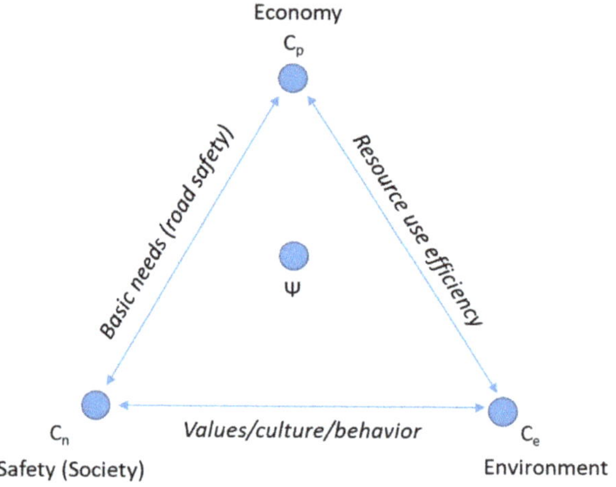

Figure 1. Sustainable development of a road tunnel: balance of safety, economy, and environment.

Our target function of the tunnel system Ψ seeks a minimum of the qualitative index ***PI***

$$\Psi = minPI \tag{1}$$

The performance Index (***PI***) has three parts

$$PI = (C_e + C_p + C_n) \tag{2}$$

where C_p is the operating costs. The decisive item is usually the price for energy for ventilation and lighting, but it also includes service costs and costs for replacement of equipment after the end of its life, labor costs of staff, and others. C_e is the cost for environmental impacts, including the cost for protection from the air pollution and others. This has also clear link to traffic performance, as congestions have clear negative impact on the produced emissions (for example CO_2 emissions [7]). C_n is the cost of traffic accidents (in general the safety impact), which consists of the material damage, cost of the emergency response, but also the cost for personal injury or death.

2.3. Urban Tunnel Integration into Smart City Concept

In terms of syntax interface between any two systems, we must focus on all seven layers of the reference ISO/OSI model [8], with the physical layer physically transmitting signals across a communication medium; the data link layer transforming a stream of bits from the physical layer into an error-free data frame for the network layer; the network layer which controls the operation of a packet transmitted from one network to another, such as how to route a packet and others. Apart from that, we must, however, also ensure that the subsystems understand each other and not only that they can exchange packages. For this we must talk about a semantic interface [9].

A semantic interface assures that the knowledge contained in each subsystem will have an unambiguous meaning for the second subsystem and form knowledge will be close to the human cognitive thinking. Ontologies [10] are suggested as a tool to assure understanding among particular subsystems.

The semantics is of a great importance, while it is often neglected. There are many different stakeholders and actors that plays a role when designing a city subsystem. Each of them has a certain perspective on the given problem. Transport engineers are interested in the traffic flow, traffic occupancy, and a number of accidents. An architect understands the city tunnel from the urbanist point of view and looks at the urbanist details. A sociologist observes many details about drivers' behavior. A business specialist looks at the operational cost, energy consumption, etc. Environmental experts study the effects of city tunnel operation on various pollution indicators.

These are often conflicting strategies of the heterogeneous fields, although we are looking at a single object (i.e., the urban tunnel). Each of these fields has a historically developed control concept according to its criteria functions, such as tunnel ventilation in relation to safety parameters, traffic flow control using variable signs, electricity control in relation to tunnel lighting, etc.

In our view, the "system of systems control" means the maximum use of the control strategies of each system (i.e., transport, energy, environment, etc.) and the creation of the best possible combination of them so that these strategies support each other and are not in conflict. This is supported for example by the work of Luin and Peteling [11]. The authors use the existing Supervisory Control and Data Acquisition (SCADA) system in connection to traffic simulation model, simulation of environmental characteristics and ventilation system. This resulting model connecting all above mentioned subsystems is used as a training base for simulating the impact of traffic accidents jointly on all subsystems and allows testing various human-machine interfaces.

2.4. Urban Tunnel Integrated Control

Let us observe M parallel information variables $(O_1, O_2, \ldots, O_M)$ of a real system (city tunnel plus its surroundings). With respect to the system complexity, it is possible to identify j-th specialization that represents typically one of following areas: transportation, energy, environment, business, etc. Within j-th specialization (in our case one of the systems) the experts can find the set of L different scenarios $(S_{j,1}, S_{j,2}, \ldots, S_{j,L})$. Scenarios $(S_{j,1}, S_{j,2}, \ldots, S_{j,L})$ must be prepared beforehand. With the help of selected scenario $S_{j,i}$, the control signals for each scenario are used to optimize the real system behavior.

We suppose that there exists an inner N- knowledge vector $(K_1, K_2, \ldots, K_N)$ common for all specializations that best characterizes strong features of the real system of systems. All scenarios $S_{j,i}(K_1, K_2, \ldots, K_N)$ are supposed to be dependent on this knowledge vector. At the same time the quality assessment function $Q_{i,j}\{S_{i,j}(K_1, K_2, \ldots, K_N)\}$ s assigned to each scenario $S_{j,i}(K_1, K_2, \ldots, K_N)$ is also dependent on this vector.

The methodology of scenarios co-ordination can include the following 5 steps:

1. Identification of knowledge base vector $(K_1, K_2, \ldots, K_N)$ from available data $(O_1, O_2, \ldots O_M)$. In transportation we can use strategic traffic detectors, in environmental areas selected climatic sensors, etc. The knowledge vector must be a low-dimensional one and all available data must be aggregated into it.
2. Selection of scenario $S_{j,i}(K_1, K_2, \ldots, K_N)$ in accordance with knowledge base $(K_1, K_2, \ldots, K_N)$. In transportation, for example, the best suited scenario is selected based on e.g., set of strategic detectors.
3. System control according to the selected scenario using real data. In transportation, for example, the selected scenario provides the traffic control of city tunnel based on measured traffic data. Appropriate control signals are distributed to variable signs. The scenarios are typically evaluated through micro/macro traffic simulations.
4. Quality assessment function depending on the selected i-th scenario in j-th specialization $S_{j,i}$ $(K_1, K_2, \ldots, K_N)$. In transportation, for example, the normalized average

travel time or the congestion length, etc. can be used as the control assessment function.

5. Cross-disciplinary i-th scenario selection for each j-th specialization to achieve the most appropriate (weighted) sum of quality assessment functions $\sum_{j,i} Q_{i,j}\{S_{i,j}(K_1, K_2, \ldots, K_N)\}$. The transportation scenario is selected, for example, to be as environmentally friendly as possible to take care of both transport and environment specializations.

The above methodology describes the most general example parametrized by vector $(K_1, K_2, \ldots, K_N)$. In the real praxis we are not concerned about (j, i)-th scenario definition, the design of scenario's selection, data collection, transmission and processing (control system). The long term knowledge of each specialization has solved this problem and we can only register how many scenarios $S_{j,i}$ exist in different areas such as transportation, energy supply, environmental protection and others [12].

Further, we suppose that it is possible to determine the (normalized) assessment function $Q_{j,i}\{S_{j,i}\}$ where (j, i)-th pair means the selected i-scenario for j-th specialization $j \in \{1, 2, \ldots\}$. If the problem has been linear we would have selected such (j, i)-th combination that simply maximizes the sum $\sum_j Q_{i,j}\{S_{j,i}\}$.

2.5. Urban Tunnel Management as a Smart City Component

This section presents an illustrative example of how the intelligent tunnel can be integrated into smart urban area. Let us assume that the tunnel communicates with whole district and negotiate its control strategy not only locally but also with respect to criterial functions of whole district.

For simplicity, let us build as a criterion function Q_j of j-sectors their financial impacts on the whole area. For example, transportation area around the tunnel can be represented by Q_1 that includes all the financial implications of the loss of drivers' time. On other side Q_2 captures the financial impacts of transport on the environment, e.g., pollution CO_2, NOX, PMx. Q_3 describes energy losses in the whole area expressed in financial parameters, including fuel consumed by all vehicles in congestion or energy consumption of own intelligent tunnel operation (ventilation, lighting).

For simplicity, consider only two possible tunnel control scenarios—either the tunnel is open S_1 or closed S_2. The main advantage of the intelligent tunnel as a smart city component is that the decision on the control strategy (S_1 or S_2) is not made locally, but with regard to the financial impact of the whole district area. Described mathematically, control is carried out using two criterion functions:

$$Q(S_1) = Q_1(S_1) + Q_2(S_1) + Q_3(S_1) \tag{3}$$

$$Q(S_2) = Q_1(S_2) + Q_2(S_2) + Q_3(S_2) \tag{4}$$

Determining the financial impacts on the whole district area is a difficult task. It is important to use all available information and prepare a traffic simulation model to allow the expression of these criterial parameters $Q(S_1)$, $Q(S_2)$ depending on the control strategies S_1, S_2.

It is obvious that the described approach can be further extended to several different control strategies, including e.g., speed limitation in the tunnel (50 or 70 km/h), ramp metering with different options of vehicles entry into the tunnel, different tunnel ventilation scenarios, etc. All these variants represent an extended set of possible control strategies $S_1, S_2, S_3, \ldots$. For each control strategy it is necessary to estimate financial impacts to the whole area to be able to provide right decision-making.

2.6. Case Study—Management of Urban Tunnel Blanka in Prague

For the purpose of city tunnel control, these three functions Q_1, Q_2, Q_3 can be expressed as monetary values. It is rather straightforward for energy management (there is a given price of energy). It is more complicated for environmental parameters and transportation, but it was addressed for example by a project GLOMODO (global model of traffic) (https://starfos.tacr.cz/cs/project/UH0370, accessed on 1 November 2021). This project aims to add values (meaning) to the data from all strategic detectors and all section speed control systems in Prague. An illustrative overview of the sensors used in the project is provided in Figure 2.

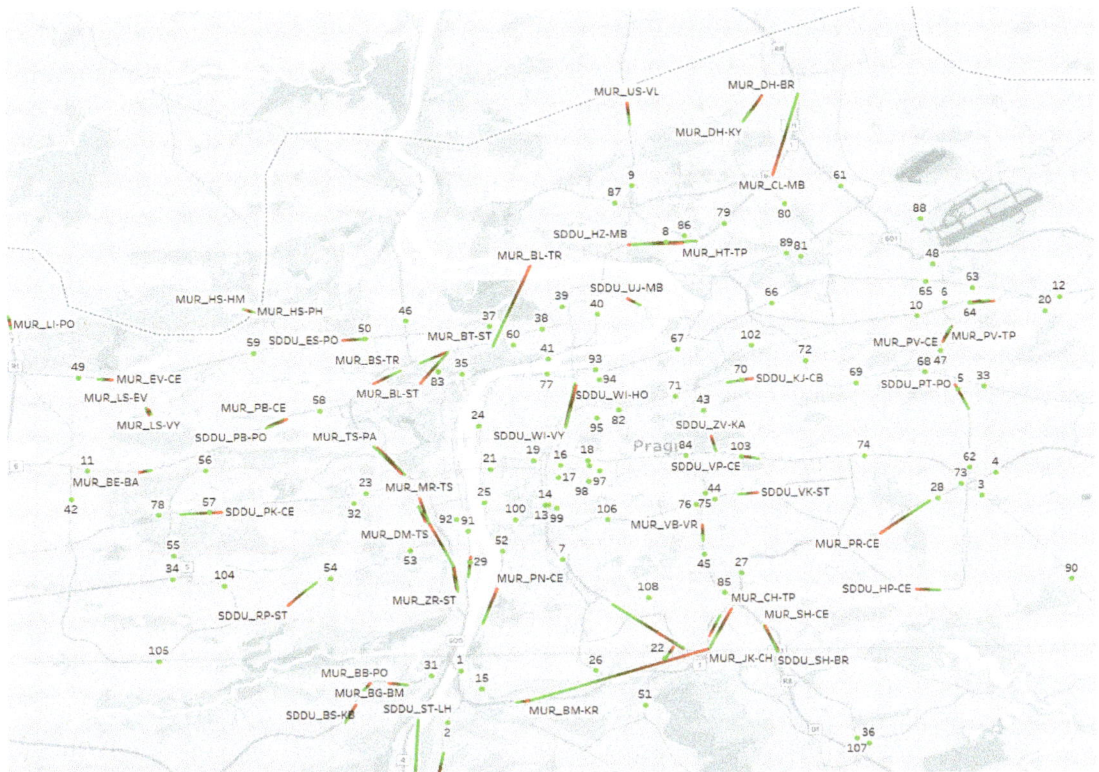

Figure 2. Overview of strategic sensors located in Prague and used within the project GLOMODO.

Project GLOMODO consists of several modules whose aim is to give a road operator or a city manager better understanding of the traffic situation. As depicted in Figure 3, apart from data preprocessing aiming mainly on identification and correction of missing data and outliers, there are three modules: Module FIS computing the cost of congestions; Module LOS providing a level of service indication based on so-called speed index [13]; and Module MEN aiming on identification normal behavior of traffic flow and warning the operators in case of any non-standard traffic patterns.

For the purpose of this paper, we shall focus on the module FIS only as it assigns monetary value based on the level of congestions. The overall cost of congestion in all district Q is computed when the vehicle is forced to travel with a limited speed due to congestions and is based on the following formula with parameters adjusted for the conditions of the city district—it is expressed by symbol (.):

$$Q(.) = Q_1(.) + Q_2(.) + Q_3(.) \tag{5}$$

where $Q_1(.)$ computes the value of lost time, $Q_2(.)$ determines the cost of environmental impact (related to increased pollutions) of congestions and $Q_3(.)$ denotes the increased fuel consumption. $Q_1(.)$ is computed as:

$$Q_1(.) = N_{act}.t_{act}.C_T.veh_{occ} \tag{6}$$

where N_{act} denotes the 5-minute value of traffic flow, t_{act} denotes the time delay due to the congested traffic, C_T the time spent in traffic and veh_{occ} the average occupancy of a vehicle. $Q_2(.)$ is computed as:

$$Q_2(.) = (S_{Gas}.C_{Gas} + (S_{Diesel} + S_{HDV}).\ C_{Diesel}).P_{CO2} \tag{7}$$

where S_{Gas} is the computed fuel consumption for petrol vehicles in the given time interval; S_{Diesel} denotes the computed fuel consumption for diesel vehicles; S_{HDV} similarly denotes the consumption of trucks and lorries; C_{Gas} denotes CO_2 emissions for 1 litre of petrol; C_{Diesel} denotes emissions CO_2 for 1 litre of diesel; and, finally, P_{CO2} is the price associated with the production of 1 ton CO_2 emissions.

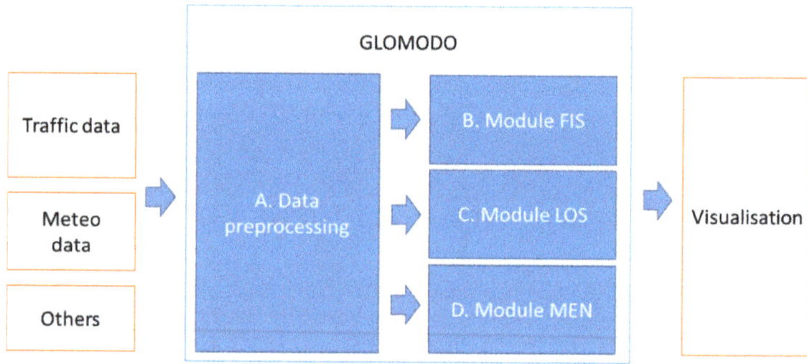

Figure 3. Main modules developed within the project GLOMODO.

$Q_3(.)$ is building upon the generally accepted models of fuel consumption for different speeds. Different tools use slightly different models, as depicted in Figure 4. For this reason, GLOMODO uses an adopted model that is based on experimental evaluations [14].

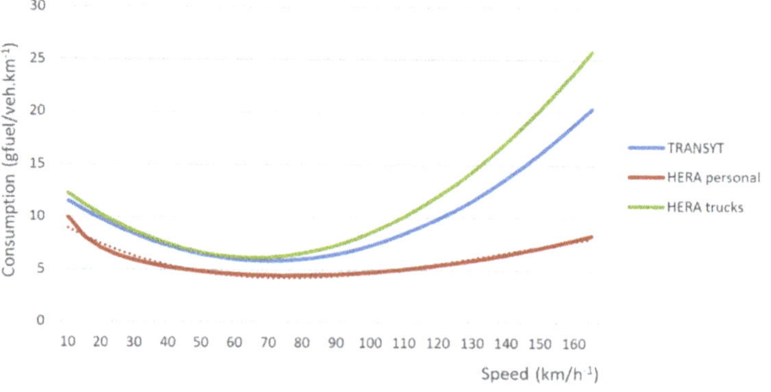

Figure 4. Comparison of fuel consumption (in liters per 100 km) according to models HERA COPERT and TRANSYT.

All parameters in the equations are set based on national as well as international guidelines and experiments. This approach is rather general and can be easily adopted for example to changes in vehicle fleets.

Examples of the results of the above-mentioned model are provided in Figure 5 where the cumulative cost of congestions on particular strategic sensors as well as the level of service (LOS) is shown.

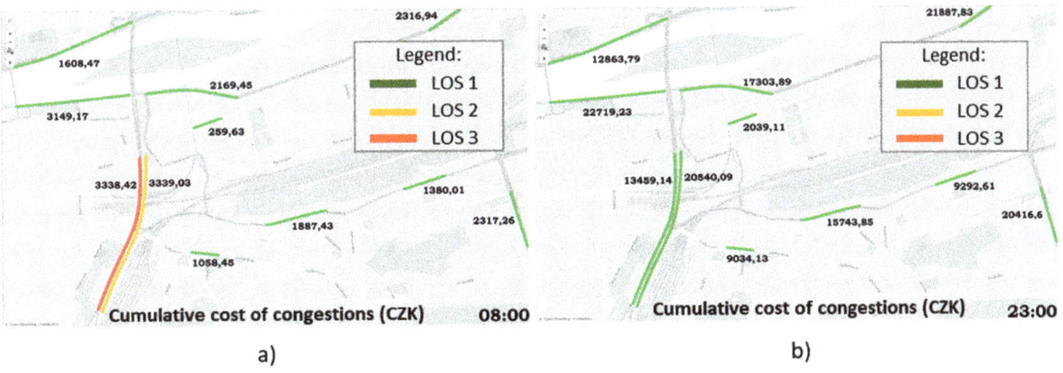

Figure 5. The cumulative cost of congestions (in CZK) and the level of service (LOS 1—green; LOS 2—yellow; LOS 3—red) as computed for an example network in Prague at (**a**) 8 a.m. and (**b**) 11 p.m.

Figure 6 further demonstrates the impact of each of the factors. Clearly, $Q_1(.)$ (i.e., the lost time) is the biggest contributor to the cost of congestion.

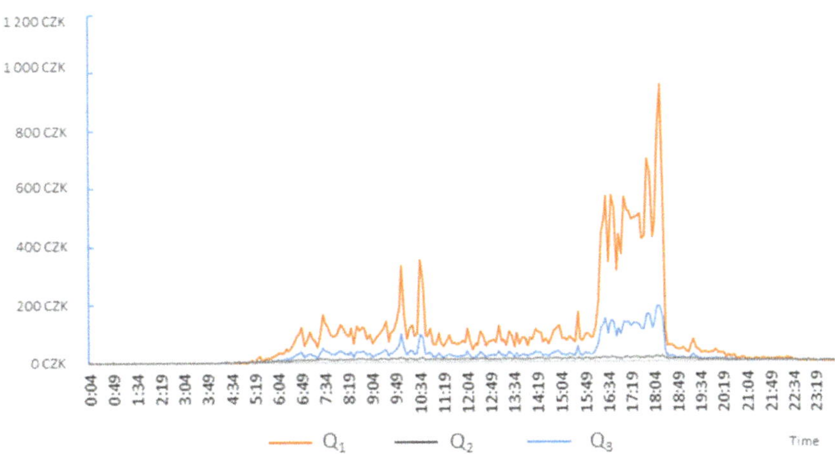

Figure 6. An example demonstrating the split of the cost of congestions (in CZK) to particular contributing factors Q_i during the day.

3. New Road Tunnel Traffic Control Algorithm

Urban road tunnels, when left without control, often face congestions. Congestions influence several aspects from performance index in Equation (2). Clearly, they impact the transport performance (less vehicles can get through the network), but they also negatively influence the environmental aspects.

The congestions in tunnels can be addressed mainly by two control approaches. Both of them are based on highway management strategies [15,16]:

A. Influencing characteristics of the traffic flow, usually through speed harmonization. It has been demonstrated [17] the capacity of a communication is increased in case a speed limit is imposed on highways. This is achieved through harmonizing the traffic flow (i.e., minimizing the speed differences among vehicles), thus decreasing the vehicular headways and thus increasing the throughput. Nowadays, this control approach is however limited by the existing norms (for example Czech norm 73 6101) stating that the tunnel is just a continuation of the road network and should maintain the same speeds. For this reason, a speed variation is not a suitable option.
B. Changing the travel demand parameters, particularly by managing (constraining) the number of vehicles entering the tunnel. This is similar to the concept of ramp metering, often adopted on highways and freeways.

Nowadays, tunnels are typically controlled by the second approach. It is, however, usually performed manually and often in a binary matter. In case queues are forming in a tunnel, the police decide to close the tunnel. This principle is demonstrated on a case study of an urban road tunnel complex Blanka (Prague, Czech republic), which is depicted in Figure 7. It is an important part of the (still not finished) Prague city ring and it was built to lighten the traffic in the central historical areas of the city. The tunnel complex is 5.5 km long and it is composed by three underground tunnels (red lines in the Figure 7). The opening of the tunnel complex Blanka in 2015 had a major impact on traffic in the city and clearly demonstrated need to take such tunnel complex as a part of the city.

Figure 7. Schematic overview of the tunnel complex Blanka.

The effect of tunnel closing clearly leads to an improvement within the tunnel, but has a strong negative impact on the traffic around the tunnel, in the rest of the city. This is described by Faltus and Hrdina in [18] and demonstrated in Figure 8. The green points denote sensors for which the average speed increased after opening the tunnel complex Blanka; the red points mean decrease in average speed. Always, the change in percent is presented. The full circle denotes the sensor that measures in one direction only, the split circles show values in both directions.

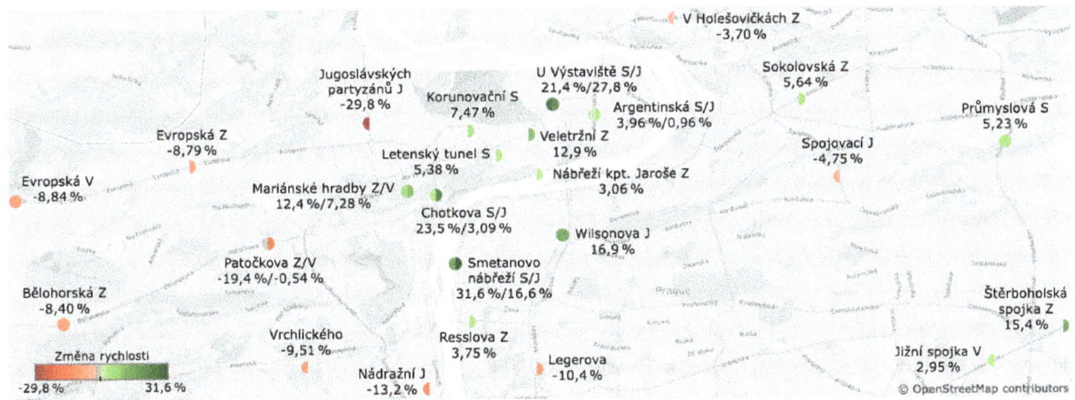

Figure 8. Changes in the average speed on selected sensors after opening the tunnel complex Blanka.

Another negative aspect of the current traffic control is the recovery phase, after the tunnel is opened again. It can take several dozens of minutes (sometimes even hours) to recover the stable state after the tunnel is opened [19].

3.1. Fundamental Traffic Flow Theory

In order to develop an optimized traffic control strategy, it is necessary to look at the characteristics of traffic flow. The traffic flow is characterized by three basic variables [20]: flow rate q, density k and mean speed u. Due to the fundamental relation $q = k.u$ there are only two independent variables. The relationship among the variables is often depicted using so-called fundamental diagrams (see Figure 9).

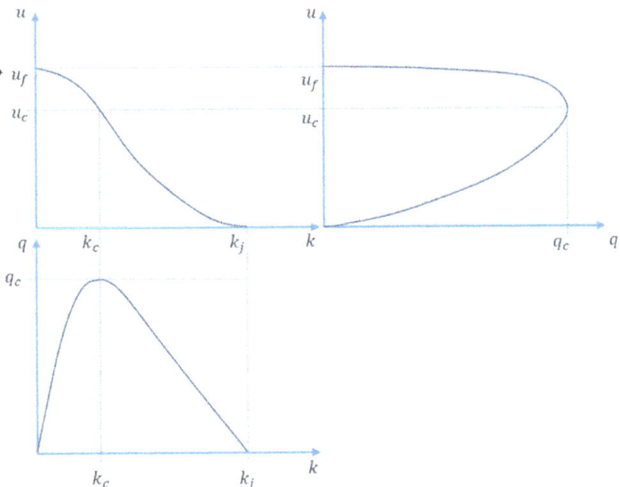

Figure 9. The fundamental diagram.

For the purpose of this paper, we need to focus on the following points and areas of interest in the diagrams:

Completely free flowing traffic, when vehicles are not impeded by other traffic and travel at a maximum (or desired) speed of u_f (free speed). This free speed depends on the road design speed (based on curvature, transverse slope, road surface quality and other parameters), the speed restrictions (speed limits), the weather, but also on the driver's characteristics. At free speed, flow rate and density will be close to zero.

Saturated traffic, when the flow rate, q, and speed, u, are down to zero. The vehicles are queuing and there is a maximum density of $\boldsymbol{k_j}$ (jam density).

Capacity traffic, with the capacity of a road equal to the maximum flow rate, $\boldsymbol{q_c}$. The maximum flow rate of $\boldsymbol{q_c}$ has an associated capacity speed, $\boldsymbol{u_c}$, and a capacity density, $\boldsymbol{k_c}$. The diagram shows that the capacity speed, $\boldsymbol{u_c}$, lies below the maximum speed, $\boldsymbol{u_f}$.

There exist various models trying to find an analytical solution fitting as best as possible the empirical measured data, for example linear model [21], triangular diagram [22], or more realistic high-order macroscopic models addressing certain discontinuity [23].

Here, we focus on the last class of models as they well describe the phenomena we face in tunnels. Ning Wu in [24] uses an approach of homogenizing traffic and addressing the transition among them. As Wu demonstrated in his paper, the measured data confirm two major data clusters (traffic states) with different characteristics as well as transition phase between them. They are called fluid traffic and jam traffic. This is confirmed also by the data measured in the Czech Republic as demonstrated for q-v relationship on example in Figure 10. This is in a way similar to Kerner's Three-phase theory [23], where next to the free flow, the jam is further split into two phases, namely synchronized flow and wide moving jam.

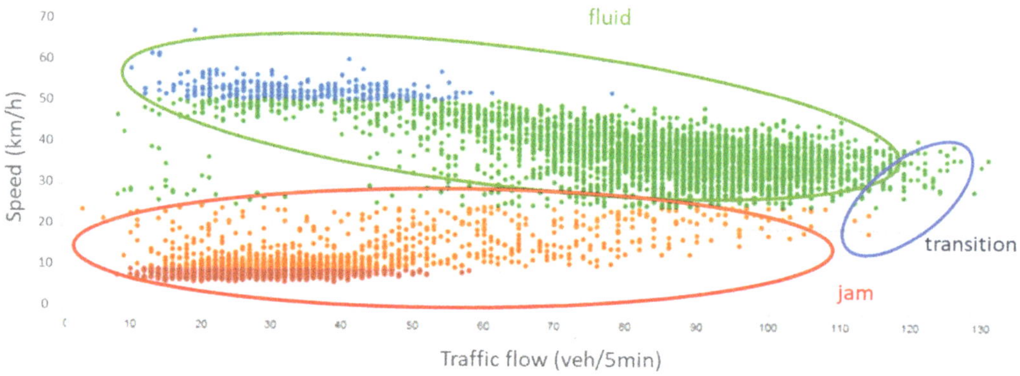

Figure 10. Data measured on Strakonicka street in Prague (October 2015) demonstrating the fluid and jammed traffic states.

The critical part which needs an extra focus is the transition from free flow to the jam and back. The traditional models often fail as there is only a limited number of measurements. This is due to the fact that the transition is rather quick (i.e., the flow does not stay in this phase for too long) and there is a clear observable hysteresis (i.e., the recovery from the jam does not follow the same trajectory as falling from the free flow to the jam). The traditional models thus tend to overestimate the road capacity. The principle of this transition is further depicted in Figure 11.

For the purpose of this paper and the road urban tunnel control system, it is clear from the above-mentioned theory that we need to focus on density. We are mainly interested in the point when the fluid traffic breaks into the jam, that is at density κ_{KO} (i.e., at the state that the average length of the net headways between the vehicles is equal to τ_{KO} in the fluid convoy). The jam traffic turns back upwards into the fluid traffic at $\kappa_{go,\,min}$ (i.e., at the state that the average length of the net headways between the vehicles is equal to τ_{go} in the jam convoy). For more details on identifying the critical values of density, refer to Wu [24].

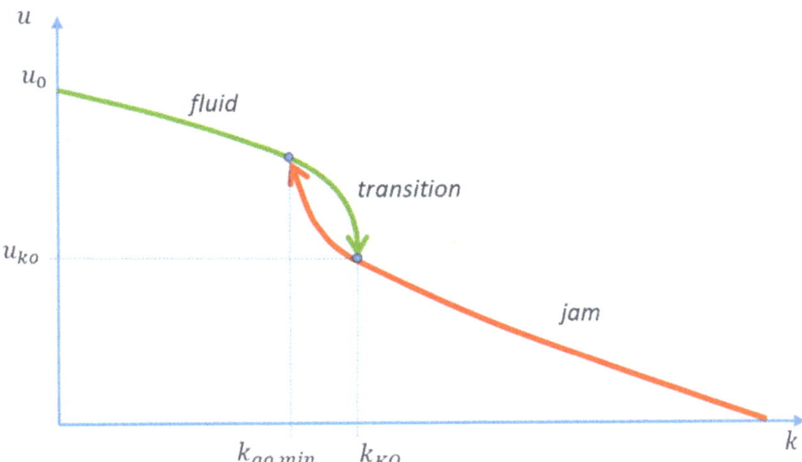

Figure 11. The hysteresis within transition phase between fluid and jam traffic.

3.2. New Road Urban Tunnel Traffic Control

As mentioned above, nowadays the most common "control" approach in an urban tunnel is a closure of a tunnel. Tympakianaki et al. [16] calls closure a disruptive traffic management strategy and further claims that the resulting strategy shall depend on the cause and type of congestions. Tunnel closure solves the problems of hygienic (environmental) limits in the tunnel. However, it means a big intervention in the quality of traffic in the whole city, which is contradicting the idea of smart cities. At the same time, due to the distributed nature of traffic spread in the city and implicit hysteresis of the traffic stream characteristics (as explained above), the recovery phase is very slow.

In this chapter we propose a new traffic control management system based on the properties of traffic flow described above. A flow chart (using the SW modelling tool Enterprise Architect) for the proposed control algorithm is provided in Figure 12 and explained within this section.

It aims to prevents traffic flow transiting into the jam phase. Similar principle of stabilizing main traffic stream is known from highways as ramp metering [25,26]. In ramp metering there is a main traffic flow going on the motorway, and a traffic signalization allows only a limited number of vehicles entering from a side on-ramp. In the case of a tunnel, we are using a similar principle, with the difference, that we are restraining vehicles entering the tunnel (see Figure 13). It is not carried out in a binary manner (tunnel open/closed), but using variable length of a red signal, T_R. The objective is to keep the density below the value of κ_{KO} (See Figure 11) and thus avoid falling into the jam traffic.

Let us consider a fixed cycle length of 120 s. The length of the red-light signal then varies between 0 s and 120 s (which in a 120-second-long cycle means that the tunnel is closed) depending on two factors.

The first parameter is the traffic density in the tunnel (see explanation in previous section and Figure 11). Since it is difficult to measure directly density, we use a parameter called occupancy, o_i (%) (the index i denotes i-th sensor). Occupancy is measured directly by point traffic sensors and under certain assumptions (mainly a uniform vehicle length L) it can be used to calculate density [27]:

$$o_i = (L + d).\ \kappa_i = c_K.\kappa_i \tag{8}$$

where L is the uniform vehicle length and d is the length of the sensor (e.g., inductive loop). While computing density, in order to allow better link to the theoretical section of this paper, we refer to the corresponding value of computed density in all calculations and figures.

To be able to react to the expected worsening of the situation in advance a prediction module is used to have a prediction of the density (for example, 15 min in advance, κ_i $(t + 15)$). There are several existing algorithms to predict traffic stream characteristics. Yu and Liu [28] are using spatial-temporal data fusion using traffic volume, road geometry, or even weather conditions. Laña et al. [29] use random forest regressors on traffic flow time series. Lv et al. in [30] predict traffic flow using a deep learning approach and Guo et al. in [31] use statistical hybrid Gaussian process regression. Altogether, the various research methods show rather strong prediction capabilities.

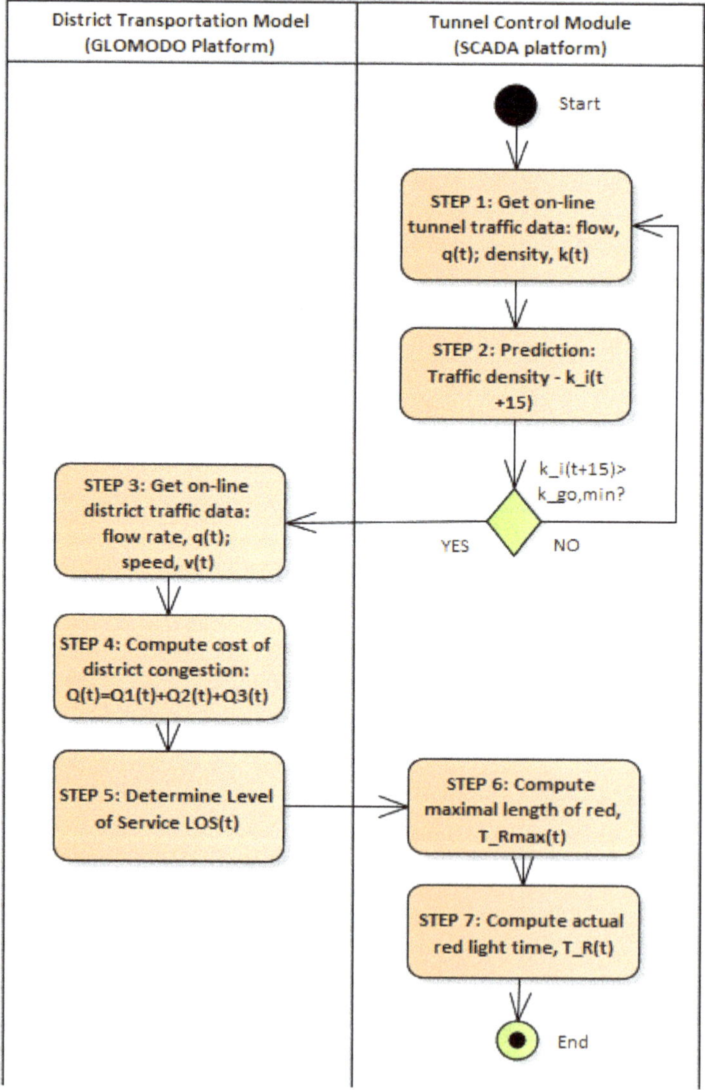

Figure 12. Flow chart for the proposed new road urban tunnel traffic control algorithm.

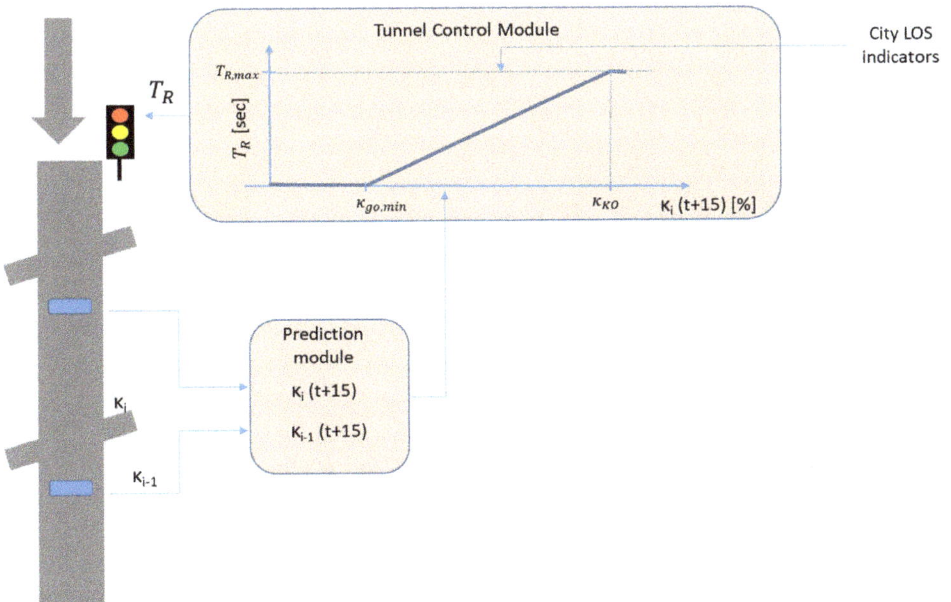

Figure 13. Schematic depiction of the road urban tunnel traffic control algorithm.

Please note that in our approach we do not wait until the density reaches value κ_{KO} (in Figure 11). We start limiting the vehicles already by $\kappa_{go,\ min}$. Clearly, we want to prevent the transition into the jam phase, so the algorithm starts earlier. This shall be sufficient as it is linked to the predicted value of occupancy.

The second parameter that influences the outcome of the traffic control is the traffic situation in the city, expressed by level of service (LOS). When the traffic situation on a selected location in the city is good, the traffic control in tunnel is not influence at all. However, if the traffic situation in the city worsens, it would be contradictory to even contribute to the worsening by improving situation in the tunnel. It can be expected that under such conditions, the situation in the tunnel worsens as well. The length of the red-light phase will be shorter.

In Figure 5 the cumulative cost of congestions and related Level of Service were introduced. The urban road traffic control algorithm works with LOS categorization—LOS 1—green; LOS 2—yellow; LOS 3—red. To obtain valuable on-line data, the system must be connected to GLOMODO (Figure 3), which continuously monitors the LOS, including available short-term predictions. This fulfills the idea of connecting the road urban tunnel with the Smart city system through presented syntactic and semantic interface.

In the case of $LOS = 1$ (green), a full interval T_R will be used for urban tunnel control according to occupancy prediction, as it is shown in Figure 13. In case of $LOS = 2$ and $LOS = 3$, the maximum of interval $T_{R,\ max}$ will be reduced accordingly. STEP 6 of the algorithm (see Figure 12) is based on the Equation (9):

$$T_{R,max}(t) = \begin{cases} MRT & for\ \ LOS\ 1 \\ 0.8 \cdot MRT & for\ \ LOS\ 2 \\ 0.4 \cdot MRT & for\ \ LOS\ 3 \end{cases} \tag{9}$$

where MRT denotes the maximum red time and it is a parameter that depends on the particular tunnel. For the case of the tunnel Blanka, the value of parameter MRT equals to

120 s (which corresponds to the cycle length discussed above). The reaction of the tunnel control system will be thus more sensitive to the surrounding traffic situation.

The actual length of the red-light signal, $T_R(t)$, i.e., STEP 7 of algorithm (see Figure 12) is currently computed according to the Equation (10):

$$T_R(t) = k_i(t+15)\cdot\frac{T_{R,max}(t)}{(\kappa_{KO} - \kappa_{go,\ min})} \tag{10}$$

As a next enhancement of the algorithm, nonlinear characteristics will be modeled using a fuzzy inference system mechanism [32,33], which will be tailored-made for specific tunnel construction and its traffic features.

3.3. Case Study Results—Control of Road Urban Tunnel Blanka in Prague

In this section, we demonstrate the effect of closing the tunnel Blanka using a BEFORE and AFTER analysis. Figure 14 shows the development of the speed index for one working day and one weekday before, resp. after the tunnel opening. The figure demonstrates that the speed index dropped more substantially before opening of the tunnel Blanka.

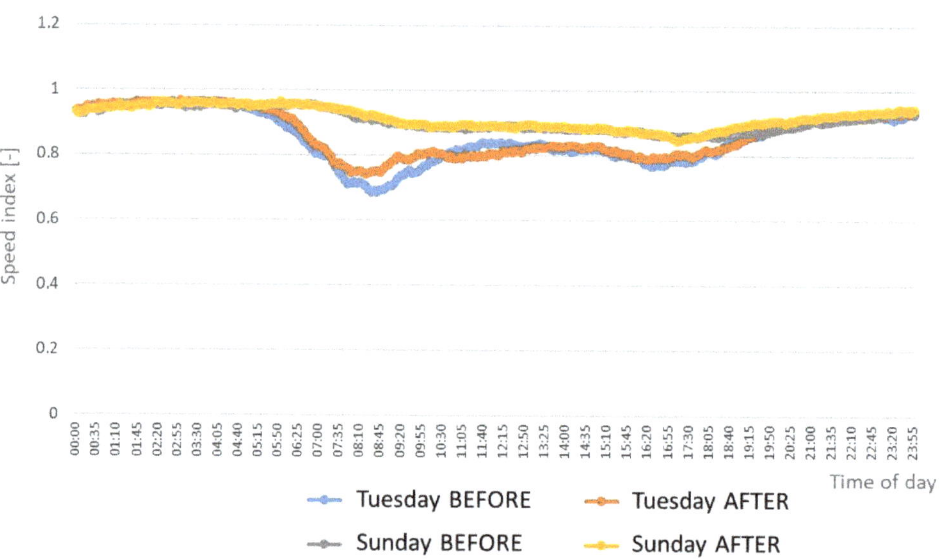

Figure 14. The average speed index over all sensors during the measurement days.

The average speed index increased from 0.909 to 0.915 for a weekend day before, respectively, after and from 0.857 to 0.865 for a selected Tuesday before, respectively, and after. The actual economic impact is demonstrated in Table 1. It uses the presented cost of congestion computations. It provides aggregated values for all traffic sensors presented in Figure 2 and always the whole day. Please note that we are looking at the impact on the rest of the city and not values in the tunnel.

The table presents the following variables:

- Sum of flow (veh/day), which takes the daily total number of vehicles on each sensor (without the traffic flow in the tunnel) and sums these numbers;
- Total cost, which provides the aggregated and cumulative cost of congestions in millions of Czech crowns;

- Index, which is just an indication of the impact of the daily sum of flows on the total price by dividing it by the variable sum of flows;
- Average speed.

Table 1. BEFORE-AFTER analysis results with respect to changes in the traffic flow (veh), total cost (mio CZK), index and average speed (km/h).

Day of Week	Phase	Sum of Flow (veh/day)	Total Cost (mio CZK)	Index: Total Cost Divided by Sum of Flow	Average Speed (km/h)
Sunday	BEFORE	3,017,261	18.97	6.29	53.8
Sunday	AFTER	2,950,674	18.35	6.22	54.9
	Change (%)	−2.21%	−3.28%	−1.10%	1.99%
Tuesday	BEFORE	4,351,693	31.50	7.24	50.5
Monday	AFTER	4,447,797	30.90	6.95	51.5
	Change (%)	2.21%	−1.89%	−4.01%	1.99%

The first part of the table focuses on weekends. We can see that there was a decrease in the total flow outside of the tunnel (−2.21%). This can mean that since the traffic is lower, more people selected to travel through the tunnel. This leads to a decrease in the total price for congestion by about 623 thousand CZK per day.

During the week, there is not only the traffic in the tunnel, but the sum of flows within the city increased by over 2.2%. Even for this increase in traffic flow, there is a decrease in the cost of congestion of about 596 thousand CZK. In both cases we also observe an increase in the average speed and especially for the weekday decrease in the Index.

Even more substantial increases of traffic (and thus deterioration of the traffic flow characteristics such as average speed) is expected if the tunnel is closed nowadays. All of the induced traffic will play an essential role to that. The tunnel attracts daily over 80,000 vehicles that in case of closure of the tunnel must disperse into the city.

4. Discussion

An urban tunnel is often presented using the word "Smart". In most cases, it however focuses on using new technologies, such as automated incident detection [34], emergency management and lighting [35], energy efficiency [36], or, for example, predictive maintenance [37]. Our aim is to move beyond new technology to integrated control strategies.

Li et al. [38] classifies a road tunnel into so-called urban underground infrastructure. They claim (for example, with Shanghai and other world cities) it to be an important part of current and future megacities. The authors confirm our conclusions that proper integration and effective and joint management of various city subsystems is essential. The impact of proper integration on sustainability is further elaborated for example in [39,40] and on an example of urban utility tunnels also [41].

The importance of implementing the cooperative control strategies is further discussed in [42]. Haddad et al. claim that a highway management without proper integration to adjacent urban intersection controllers is not sufficient. They provided a model for the mixed environment and used model predictive control to optimize the traffic control at the intersections, ramp metering and speed harmonization. Our paper clearly focuses on a similar approach, when the adjacent city network influences the parameters of the tunnel control system. As part of our control design, we managed to convert the cooperative tunnel management into a well-known method of harmonizing traffic flow described in [43,44]. It is therefore possible to use these control results to determine the required key performance indicators (KPIs) of whole district area.

The presented results of this paper are in accordance with the ongoing fourth industrial revolution [45], the basic feature of which is the interconnection of sub-subsystems into more complex system. The importance of understanding the distributed nature of the control problem was described in [46]. An example of this trend is also the concept of smart cities and the interconnection of its elements, such as the road urban tunnel. The subsystems can no longer be controlled by standard methods based on the collection and processing of local data, but it is necessary to use artificial intelligence algorithms that can work with knowledge bases based on ontologies and perform more sophisticated decision-making and management of these complex systems [47].

In order to be able to specify the impact of certain control strategies on the entire city, we have moved beyond the traditional key performance indicators (KPIs) [48]. We offered a solution to measure the environmental and social impact of congestions by its monetary equivalent. This allows the control strategies to take into account integration of various subsystems in a distributed environment.

In this paper, we focused on an example of how to approach the management of a road urban tunnel with knowledge of the traffic situation in its neighborhood. The above results show that even if the tunnel is controlled so that the passage of vehicles is minimal, it is far better than if it was closed, as is the case today. Although its closure solves the local situation inside the tunnel, it significantly worsens the traffic in the surrounding district. If other elements of transport infrastructure are managed by similar methods, it is possible to expect better transport throughput and the achievement of a better quality of life in these cities.

5. Conclusions

Within this paper, a system analysis of both the tunnel as a complex system and the method of its integration into the smart city concept was presented. The result of the analysis yields to the conclusion that the road urban tunnel integration into the concept of smart city can bring many benefits to municipality and also to drivers.

The level of service (LOS) was introduced, which represents the transport quality in the whole district based on quantification of congestion prices, namely financial implication of the loss of drivers' time, financial impacts of transport on the environment and energy losses in the whole area. The required parameters were obtained for Blanka tunnel in Prague thanks to the described GLOMODO (global model of traffic in the capital Prague) system.

The current practice of many cities, which include a road urban tunnel, deals with tunnel control by closing the tunnel in the case of congestion to prevent congestion inside the tunnel. It is a logical and local solution, since congestion in the tunnel is expensive in terms of ventilation and other support services. On the other hand, it will significantly worsen congestion outside the tunnel itself.

For these reasons, a new "road urban tunnel control algorithm" has been proposed, which would not only contain two states "open" and "closed", but which would let vehicles enter into the tunnel in small amount and maintain the traffic flow within the transition phase between fluid and jam traffic. Experiments performed in the Blanka tunnel in Prague have shown that even such a small traffic flow can significantly contribute to solving a complex traffic situation in the entire area.

In the future, we intend to develop more advanced road urban tunnel control algorithm, based on the principle of fuzzy methods, which allows the control strategy to be adapted to the specific construction and traffic parameters of a given tunnel.

The results presented here were demonstrating the working principle of the algorithm. In order to have more clear view on the real impact of the method on tunnel as well as city parameters, a large-scale study will be conducted. Here also the environmental aspects shall be measured.

The paper can be understood as a methodology for the integration of similar infrastructural elements (bridges, railway crossing, etc.) and their management into a complex system for a smart city. Thanks to the knowledge of a wider area (neighborhood, district, etc.) than the element itself, it is possible to achieve better sustainability of a specific part of the city and to increase resilience of the whole solution.

Author Contributions: Conceptualization, O.P., P.P. and M.S.; methodology, O.P., P.P. and M.S.; validation, M.S.; writing—original draft preparation, P.P., O.P. and M.S.; writing—review and editing, O.P. All authors have read and agreed to the published version of the manuscript.

Funding: This research was funded by the European Regional Development Fund under the project AI & Reasoning (reg. No. CZ.02.1.01/0.0/0.0/15_003/0000466).

Institutional Review Board Statement: Not applicable.

Informed Consent Statement: Not applicable.

Conflicts of Interest: The authors declare no conflict of interest.

References

1. Ruhlandt, R.W.S. The governance of smart cities: A systematic literature review. *Cities* **2018**, *81*, 1–23. [CrossRef]
2. Schmitz, P. PIARC Road Tunnels Manual. PIARC. Version 2019. Available online: https://tunnelsmanual.piarc.org/en (accessed on 1 November 2021).
3. Rzevski, G.; Skobelev, P. *Managing Complexity;* WIT Press: Southampton, UK, 2014.
4. Grösser, S.N. Complexity management and system dynamics thinking. In *Dynamics of Long-Life Assets;* Springer: Singapore, 2017; pp. 69–92.
5. PIARC. Technical committee 3.3 road tunnel operation. In Proceedings of the Road Tunnels: Complex Underground Road Networks (2016R19EN), Da Nang, Vietnam, 23–25 October 2013; ISBN 978-2-84060-404-4.
6. Flint, R.W. Basics of Sustainable Development. In *Practice of Sustainable Community Development;* Springer: Singapore, 2012; pp. 25–54.
7. Dong, Y.; Xu, J.; Liu, X.; Gao, C.; Ru, H.; Duan, Z. Carbon emissions and expressway traffic flow patterns in China. *Sustainability* **2019**, *11*, 2824. [CrossRef]
8. Fraihat, A. Computer networking layers based on the OSI model. *Test Eng. Manag.* **2021**, *83*, 6485–6495.
9. Svítek, M.; Votruba, Z.; Moos, P. Towards information circuits. *Neural Netw. World* **2010**, *20*, 241–247.
10. Přibyl, P.; Přibyl, O.; Svítek, M.; Janota, A. Smart city design based on an ontological knowledge system. In *Communications in Computer and Information Science;* Springer: Singapore, 2020; Volume 1289, pp. 152–164.
11. Luin, B.; Petelin, S. Coupling models of road tunnel traffic, ventilation and evacuation. *Transport* **2020**, *35*, 336–346. [CrossRef]
12. Svítek, M. *Information Physics-Physics-Information Analogies for Complex Systems Modelling;* Elsevier: Amsterdam, The Netherlands, 2021; ISBN 978-0-323-91011-8.
13. He, F.; Yan, X.; Liu, Y.; Ma, L. A traffic congestion assessment method for urban road networks based on speed performance index. *Procedia Eng.* **2016**, *137*, 425–433. [CrossRef]
14. Přibyl, P.; Přibyl, O.; Faltus, V.; Hrdina, L.; Matowicki, M. Methodology for determining price of congestions in Prague (in Czech-Metodika Kvantitativního hodnocení kongescii V HL. M. PRAZE). *Res. Rep.* **2019**, 51.
15. Pribyl, O.; Koukol, M.; Kuklová, J. Computational intelligence in highway management: A review. *Promet-Traffic Transp.* **2015**, *27*, 439–450. [CrossRef]
16. Tympakianaki, A.; Koutsopoulos, H.N.; Jenelius, E. Anatomy of tunnel congestion: Causes and implications for tunnel traffic management. *Tunn. Undergr. Space Technol.* **2019**, *83*, 498–508. [CrossRef]
17. Beneš, J.; Přibyl, O. Effects of highway management on traffic flow characteristics. *Arch. Transp. Syst. Telemat. 7* **2014**, *2*, 14–18.
18. Faltus, V.; Hrdina, L. Systémové modelování přínosů tunelu Blanka. *Silniční Obz.* **2018**, *79*, 63–69. (In Czech)
19. Helbing, D.; Treiber, M. Critical discussion of synchronized flow. *Cooper. Transp. Dynam.* **2002**, *1*, 2.1–2.24.
20. Kühne, R. Foundation of traffic flow theory I: Greenshields legacy highway traffic. In *Symposium on the Fundamental Diagram: 75 Years (Greenshields 75 Symposium) Transportation Research Board;* Woods Hole: Falmouth, MA, USA, 2008.
21. Greenshields, B.D. A study of highway capacity. *Highw. Res. Rec.* **1935**, *14*, 448–477.
22. Yperman, I.; Logghe, S.; Immers, B. The link transmission model: An efficient implementation of the kinematic wave theory in traffic networks. In Proceedings of the 10th EWGT Meeting and 16th Mini-Euro Conference, Poznan, Poland, 13–16 September 2005.
23. Kerner, B. Three-phase traffic theory and highway capacity. *Phys. A Stat. Mech. Appl.* **2004**, *333*, 379–440. [CrossRef]
24. Wu, N. A new approach for modeling of Fundamental Diagrams. *Transp. Res. Part. Policy Pr.* **2002**, *36*, 867–884. [CrossRef]
25. Gaffney, J.; Zurlinden, H. City Wide Coordinated Ramp Meters-Keeping Urban Motorways Safe and Efficient With City Wide Coordinated Ramp Meters. *Int. Encycl. Transp.* **2021**, 10–20. [CrossRef]
26. Papageorgiou, M. Overview of road traffic control strategies. *IFAC Proc.* **2004**, *37*, 29–40. [CrossRef]
27. Gartner, N.; Messer, C.J.; Rathi, A.K. *Traffic Flow Theory-A State-of-the-Art Report: Revised Monograph;* Turner-Fairbank Highway Research Center: McLean, VA, USA, 2002; p. 386.
28. Yu, G.; Liu, J. A hybrid prediction approach for road tunnel traffic based on spatial-temporary data fusion. *Appl. Intell.* **2019**, *49*, 1421–1436. [CrossRef]
29. Lana, I.; Olabarrieta, I.I.; Del Ser, J.; Rodriguez, L. Data-driven predictive modeling of traffic and air flow for the improved efficiency of tunnel ventilation systems. In Proceedings of the 2020 IEEE 23rd International Conference on Intelligent Transportation Systems (ITSC), Rhodes, Greece, 20–23 September 2020; pp. 1–6.
30. Lv, Y.; Duan, Y.; Kang, W.; Li, Z.; Wang, F.-Y. Traffic flow prediction with big data: A deep learning approach. *IEEE Trans. Intell. Transp. Syst.* **2014**, *16*, 865–873. [CrossRef]

31. Guo, J.; Chen, F.; Xu, C. Traffic flow forecasting for road tunnel using PSO-GPR algorithm with combined kernel function. *Math. Probl. Eng.* **2017**, *2017*, 2090783. [CrossRef]
32. Balal, E.; Cheu, R.L. Comparative evaluation of fuzzy inference system, support vector machine and multilayer feed-forward neural network in making discretionary lane changing decisions. *Neural Netw. World* **2018**, *28*, 361–378. [CrossRef]
33. Koukol, M.; Přibyl, O. Fuzzy algorithm for highway speed Harmonisation in VISSIM. In *Robotics*; Springer: Singapore, 2013; pp. 459–467.
34. Fed4sae (Federated Cps Digital Innovation Hubs for the Smart Anything Everywhere Initiative) Website. Available online: https://fed4sae.eu/innovative-projects/smart-tunnel/ (accessed on 15 August 2021).
35. Smartcitystreets Website. Available online: https://smartcitystreets.com/tunnels/ (accessed on 15 August 2021).
36. Schreder Website. Available online: https://www.schreder.com/en/projects/smart-tunnel-lighting-solution-ensures-safety-comfort-velser-tunnel (accessed on 15 August 2021).
37. Smart Tunnel Project Website. Available online: https://wtc2023.gr/smart-tunnel-project/ (accessed on 15 August 2021).
38. Li, H.Q.; Fan, Y.Q.; Yu, M.J. Deep Shanghai project–A strategy of infrastructure integration for megacities. *Tunn. Undergr. Space Technol.* **2018**, *81*, 547–567. [CrossRef]
39. von der Tann, L.; Ritter, S.; Hale, S.; Langford, J.; Salazar, S. From urban underground space (UUS) to sustainable underground urbanism (SUU): Shifting the focus in urban underground scholarship. *Land Use Policy* **2021**, *109*, 105650. [CrossRef]
40. Translator Qiao, Y.-K.; Peng, F.-L.; Sabri, S.; Rajabifard, A. Socio-environmental costs of underground space use for urban sustainability. *Sustain. Cities Soc.* **2019**, *51*, 101757. [CrossRef]
41. Shahrour, I.; Bian, H.; Xie, X.; Zhang, Z. Use of smart technology to improve management of utility tunnels. *Appl. Sci.* **2020**, *10*, 711. [CrossRef]
42. Haddad, J.; Ramezani, M.; Geroliminis, N. Cooperative traffic control of a mixed network with two urban regions and a freeway. *Transp. Res. Part. B Methodol.* **2013**, *54*, 17–36. [CrossRef]
43. Diakaki, P.; Kotsialos, D.W. *Review of Road Traffic Control Strategies*; IEEE: Piscataway, NJ, USA, 2003; Volume 91, pp. 2041–2042.
44. Přibyl, P.; Přibyl, O.; Apeltauer, T. Mobile highway telematics system on D1 highway in the Czech Republic. *Proc. IEEE* **2014**, *91*, 386–395.
45. Lom, M.; Přibyl, O.; Svítek, M. Industry 4.0 as a Part of Smart Cities. In *2016 Smart Cities Symposium Prague (SCSP)*; IEEE: Piscataway, NJ, USA, 2016; ISBN 978-1-5090-1116-2.
46. Mustapha, K.; Mcheick, H.; Mellouli, S. Smart cities and resilience plans: A multi-agent based simulation for extreme event rescuing. In *Public Administration and Information Technology*; Springer: Singapore, 2016; pp. 149–170.
47. Janota, A.; Nemec, D.; Hruboš, M.; Pirník, R. Knowledge-Based Approach to Selection of Weight-in-Motion Equipment. *Commun. Comput. Inf. Sci.* **2016**, *640*, 1–12. [CrossRef]
48. Hara, M.; Nagao, T.; Hannoe, S.; Nakamura, J. New Key Performance Indicators for a Smart Sustainable City. *Sustainability* **2016**, *8*, 206. [CrossRef]

Article

A Geometrical Structure-Based New Approach for City Logistics System Planning with Cargo Bikes and Its Application for the Shopping Malls of Budapest

Dávid Lajos Sárdi * and Krisztián Bóna

Department of Material Handling and Logistics Systems, Budapest University of Technology and Economics, 1111 Budapest, Hungary; krisztian.bona@logisztika.bme.hu
* Correspondence: david.sardi@logisztika.bme.hu

Featured Application: In our paper, we developed a new, geometrical structure-based approach and its graph theory-based geometric model with graph theory-based notation, which can be used in the future for city logistics system planning with cargo bikes.

Abstract: Nowadays, cargo bikes are seeing an ever-greater role in city logistics with an increasing number of deliveries, and it is essential to examine their future role in green and smart cities. In our work, we examine the application of cargo bikes in the city logistics system of the urban concentrated sets of delivery locations, focusing first on shopping malls, with the investigation of the geometrical structure of the logistics network. In the examined concept, the use of cargo bikes will be combined with electric trucks to make possible green deliveries of urban concentrated sets of delivery locations. In this paper, we present the experiences of the existing systems and the related research, the simulation model of the examined new concept with cargo bikes and its results, the graph theory-based geometric model of the examined city logistics system with graph theory-based notation, and the application of the new approach for Budapest. The main output of this research is the geometrical model of the urban concentrated sets of delivery locations and its application. Based on this geometrical model, it will be possible to decide about the suitability of the examined cargo bike-based city logistics concepts for given cities.

Keywords: city logistics; cargo bike; cargo cycle; simulation; geometrical model; shopping mall; concentrated sets of delivery locations; bike delivery

Citation: Sárdi, D.L.; Bóna, K. A Geometrical Structure-Based New Approach for City Logistics System Planning with Cargo Bikes and Its Application for the Shopping Malls of Budapest. *Appl. Sci.* **2021**, *11*, 3300. https://doi.org/10.3390/app11083300

Academic Editor: Leon Rothkrantz

Received: 5 March 2021
Accepted: 4 April 2021
Published: 7 April 2021

1. Introduction

In the City Logistics Research Group of the Department of Material Handling and Logistics Systems at the Budapest University of Technology and Economics [1], we have focused on the logistics systems of the so-called urban concentrated sets of delivery locations (abbreviated as CSDL) since 2015. After data collection, development of several models, and complex analysis, we started a project to examine how to use cargo bikes in the new, innovative city logistics systems of shopping malls—one of the most critical CSDLs—and the first studies have shown that it is worth focusing on this area. In the introduction, the definition of the CSDLs, the purpose of the research, and the structure of the paper can be seen.

In the approach of the City Logistics Research Group, the urban delivery locations have two main groups. There are single delivery locations and concentrated sets of delivery locations (urban zones or buildings with a high density of delivery locations), which are groups of single delivery locations with different concentration degrees [2]. Based on our previous research results [3], there are two main types of concentration: concentration with either open or closed infrastructure. In the case of open concentration, the set is marked by an open area, so the stores are surrounded by roads and squares, such as in the case of a shopping or pedestrian area surrounded by urban roads or an open market surrounded by a square. Within a closed infrastructure, the CSDL is marked by a building that makes a set from the single delivery locations. These main groups of concentrated sets of delivery locations have several types (subgroups), as can be seen in Figure 1. It can be concluded that this classification and the characterization of the properties of the typical delivery locations are also significant results of this research. This classification can be important for city logistics system planning nowadays and in the smart cities of the future as well; as for the deliveries of the CSDLs different solutions are needed than in the case of the single delivery locations.

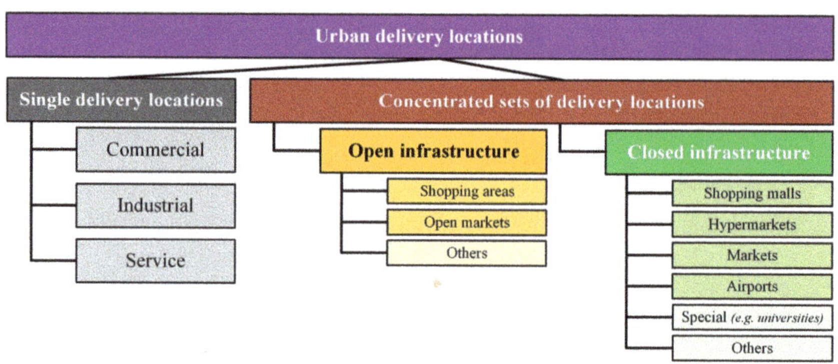

Figure 1. Groups of urban delivery locations [3].

As the first step of this research project, a complex data collection methodology was developed [4], which helped to collect data about 5 CSDLs from Budapest (4 shopping malls and 1 shopping area) and 540 stores within them. In Budapest, there are 32 shopping malls with a sum of 3432 stores. In the data collection phase, 4 of these 32 shopping malls were examined, with a sum of 663 stores (19% of all the stores from shopping malls in Budapest), and 377 of them answered our questionnaire (57% of the examined stores and 11% of all the stores of shopping malls in Budapest). Additionally, we examined the most significant shopping area of the city (the "Váci utca" shopping area), with 418 stores. In this research, 163 of them answered our questionnaire (39% of the stores of the shopping area). The collected data provided input for the modeling, where a complex mathematical model [3], a mesoscopic simulation model for the examination of the physical parameters and the cost structure [5], and a topological model for shopping areas [3] were developed. Based on the results provided by the simulation, the consolidation-based, multi-stage city logistics solutions can result in significant savings in the examined systems, both in terms of performance indicators and costs.

The field understudy is significantly important because in every former research, only some sub-systems of the systems were examined, and not with the analysis, simulation, and optimization of the whole systems with several different types of CSDLs. Much research examined the consolidation centers, the cross docks, and the technological solutions, and some projects focused on some well-defined types of CSDLs. In one project, the optimal position of shopping malls and logistics centers were examined based on performance and cost parameters [6], and in another one, the loading and parking areas of a shopping mall were in focus [7]. In some cities, there are existing consolidation-based systems in the city logistics system of (primarily) shopping areas, e.g., in Bristol, Padova, and Nijmegen [8], and in Budapest, there was such a system in the 1970s. The implementation of these solutions led to significant savings, and their results were used as inputs.

In the research of our Research Group, several innovative, gateway-concept-based city logistics solutions were developed. We examined fuel oil-powered small trucks, green trucks, and vans, the use of urban railways and cargo boats; the suitability of the metro network for freight transport was examined too, and cargo bikes had different roles in several solutions, such as deliveries, home deliveries, or deliveries between stores. In this paper, we would like to examine a concept of serving the shopping malls of Budapest using small green trucks and cargo bikes, based on the investigation of the geometrical structure of the logistics network [9]. The primary purpose of the paper is to present this new concept, which is basically a new city logistics approach, where cargo bikes are combined with electric trucks in a city logistics system of concentrated sets of delivery locations, based on the geometrical structure of the logistics network. In our paper, we show how this new approach can be examined and described. It was expected that such a solution could help to reduce emissions in the examined city logistics system and to reduce the load of the urban roads at the same time. It can be assumed, too, that such a solution can be more expensive than a simple city logistics system with direct, consolidated deliveries, but regarding this question, it is very important that in the case of any traffic issues, the cargo bikes can have better possibilities to perform the delivery tasks. To show these expected effects is an essential purpose of this paper.

For this, first, we present those existing city logistics solutions and previous research projects which provided data, ideas, and experience for our work. From these, those projects were the most important where the combined use of cargo bikes and lorries were examined. Next, a complex data analysis will be presented for the concentrated sets of delivery locations of Budapest to show the usage possibilities of cargo bikes in their city logistics system. In the case of this data analysis, 327 stores from 3 shopping malls (while in the sum 32 shopping malls of the city there are 3432 stores) and 163 stores of the biggest shopping area of Budapest (where the sum number of stores at the data collection was 418) were examined, as in their case, we had available data about the cargo bike delivery-related questions. After the data analysis, we present a macroscopic simulation model to examine this new city logistics approach in the case of all 32 shopping malls of Budapest, and the results of the simulation will show the expected effects of the examined approach with cargo bikes and electric trucks. At the simulation, the most important purpose is to get data about the expected number of delivery vehicles, about the performances, and about the expected delivery costs of the examined system. Before the simulation, it was necessary to estimate the demands of the unknown shopping malls as well; for this, an ABC-analysis-based new solution was used, which will be presented in this section, too. After these, we show the relevant urban structures and the theoretical, symmetrical, and real geometrical model of the examined city logistics system, with a graph theory-based notation and its application for the simulated city logistics system from Budapest. In the last section of our paper, we write about the next steps of the research, where the possible reference structure of the geometrical model must be highlighted, and the use of a multicriterial model will be very important as well to make possible for any examined city a decision about the use of the new approach.

The main sections of the paper are the followings:

- Materials and Methods: use of cargo bikes in city logistics systems—presentation of existing cargo bike projects, previous research experiences, data analysis for Budapest regarding the possible city logistics use of cargo bikes, the simulation model of the examined city logistics system, and results of the simulation.
- Results: the theoretical geometrical model and its application for Budapest—presentation of the related urban structures, the theoretical, the symmetrical, and the real geometrical model with graph theory-based notation and its application for Budapest.
- Discussion and future research—discussion of the results and presentation of the next steps of the research, highlighting the analysis of the reference geometrical structure and the development of a multicriterial ranking model for decision support.
- Conclusions—summary of the paper and its conclusions.

2. Materials and Methods: Use of Cargo Bikes in City Logistics Systems

Cargo bikes in city logistics systems are becoming more and more popular, and they can be useful in smart cities as well, as they are very flexible, and it is very simple to integrate them into any city logistics solution. In addition to single two-wheeled cargo bikes, bigger equipment and special trailers can be used, which are able to deliver even pallet-sized unit loads. Most of them have electric power assistance, so they can be used for bigger distances and for larger amounts. Their examination is critical, similarly to every other electric and CO_2-free solution (like electric lorries, drones, or electric cargo mopeds), as getting a CO_2-free urban freight transport is one of the main purposes of the European Union in this field [10]. Electric vehicles are in the focus of current research projects as well, but these projects are examining mostly the usual road delivery solutions [11]; their costs, their range, charging times, and their capacity is essential as well [12], and it became necessary to develop new city logistics concepts in this field. It is also clear that the number of electric vehicles in city logistics systems is increasing, and they have similar benefits as the cargo bikes, but as they have a long charging time and their range is limited, a significant drawback can be concluded in comparison with the most used urban vehicles [13]. In contrast, in the case of cargo bikes, the batteries are combined with human power, and congestion is not a problem for them, so it is important to examine them in these new city logistics concepts. Next to cargo bikes, delivery drones are examined in several research projects as they are getting more and more popular, and they can be used for more and more tasks in city logistics [14,15], but if we compare cargo bikes to them, there will be still several advantages: batteries are still combined with human power, landing is not a problem for cargo bikes, they have bigger capacity, and the flight regulations do not mean restrictions for cargo bikes. Based on these, it can be concluded that cargo bikes have several advantages if they are compared with the other electric urban delivery solutions, so it is very important to calculate with them in the case of any city logistics developments.

2.1. Existing City Logistics Systems with Cargo Bikes

Based on studies, nowadays, in most European cities, every second motorized travel in freight transport could be handled by bikes instead of passenger cars and lorries [16]. In the case of home deliveries, every fourth package could be delivered by two-wheeled utility bikes [16]. Accordingly, such systems are operating in a large number of European cities. For example, DHL works in several cities with cargo bikes. We would like to highlight their cubicycle system, a four-wheeled construction [17], which is used in several cities in the Netherlands (e.g., in Utrecht), and additionally in Gent, Frankfurt, and Taipei. In these systems, EUR-pallet-sized (EUR-pallet: 800 mm × 1200 mm) intermodal units are delivered by lorries to transshipment points where they are passed over by a special equipment to the cargo bikes (Armadillo cubicycles), and they solve the LTL (less thank truck load) last-mile deliveries of the delivery locations of their zone. In the geometrical model-based new approach, which is presented in this paper, the deliveries are organized in a similar way, with more stages and transshipment points, but the routes will be selected based on the geometrical structure of the network.

Another city logistics system that should be highlighted is Eadessopedela from Rome, Italy. In this system, from a so-called "mobile storage" the goods are loaded to cargo bikes at a higher point of the city, and they go to the delivery locations by using the advantages of the terrain [18]. This mobile storage-based concept is dynamic, so the performances and the emissions can be minimized. A similar system is operated by UPS in Hamburg, Dublin, and Leuven [19], and TNT had a similar pilot project in Brussels [20]. In both cases, an intermodal unit functioned as mobile storage, and the cargo bikes were loaded from them.

Additionally, several city logistics solutions can be mentioned: At Vanapedal, in Barcelona, two- and three-wheeled vehicles are used; at Zolle in Rome, only fruits and vegetables are delivered; and the logistics system of Outspoken in Cambridge handles last-mile deliveries and postal mails. In the KoMoDo pilot project in Berlin [21], Hermes operated with cubicycles with special trailers, so it became possible to handle two EUR-pallet-sized intermodal units in their last-mile deliveries [22].

It may also be interesting to examine projects where cargo bikes are combined with other green transportation modes. In Paris, in the Vert Chez Vous system [23], barges deliver cargo trikes on the Seine, and the trikes are loaded with small packets on the boat during the delivery. As a result of this solution, 15 trucks can be replaced each day [24]. DHL has a combined solution in Amsterdam with small cargo boats and cargo bikes, and in Frankfurt, in the LogistikTram pilot project, cargo bikes were combined with cargo trams [25]. In this solution, special logistics units were loaded to the bikes from the trams at transshipment points.

Based on the analysis of 50 city logistics projects, a study concluded that cargo bike-based city logistics solutions could lead to high profit. The authors of this paper highlighted that their result is coherent, and other studies proved the same advantages, not only from an economic point of view but also in terms of the environment and the quality of life [26].

As in our former project, from where the graph theory-based idea comes, primarily Budapest was examined, so the existing cargo bike solutions of the city are important as well. In the city, the length of the bike path-network was 187 km before the pandemic in 2020 [27], a bike-sharing system is available, and there are approximately ten logistics provider companies that use utility bikes for their deliveries. Only some of the active providers are operating cargo bikes or cargo trikes, and significant development potentials can be expected in this field. The biggest logistics provider with cargo bikes in the city is Hajtás-Pajtás [28], which is operating utility bikes, two- and three-wheeled cargo bikes, and an electric lorry, and is organizing deliveries to the city center from a micro consolidation center.

2.2. Summary of Research on Cargo Bikes

Naturally, as cargo bikes are becoming more widespread, several research projects have examined cargo bike-based city logistics systems. Most of these studies deal with their application possibilities and with the possible environmental and financial effects of the application. Based on a study from 2014, most of the available publications and documents are examining cargo bikes in Europe, and they are looking for the potential application fields of them; but it is generally argued that cargo bike solutions are a viable alternative for the future [29]. It is also interesting that cargo bikes can be considered as an alternative to walking in the case of the last section of urban deliveries between the loading area and the customer, similarly to drones [30]. A study from 2014 concluded that if the costs, payloads, and range are examined, (electric) cargo bikes are between simple bikes and passenger cars [31], which also have an important role in urban transportation and in urban logistics.

Most of the studies concluded that motorized delivery transactions should be replaced by new solutions, which is why cargo bikes are getting more attention. Based on a paper from 2012, it was primarily the smaller companies that were working with cargo bikes in the CEP-sector (Courier, Express, and Parcel), and their application was motivated mostly by the problems of urban transport and by green aspects [32]. Nowadays, most of the bigger city logistics providers (like DHL or GLS) are expanding their roles in cargo bike deliveries.

Based on a paper from 2017, approximately 10% of lorries and trucks could be replaced nowadays by cargo bikes (without decreasing efficiency) in urban areas where there is a maximum of 2 km linear distances. This could reduce CO_2 emissions significantly, and the external costs could be reduced by 25%. Based on these, cargo bikes could be real alternatives for the transport situation of overcrowded and severely polluted city centers [33]. In another paper from 2019, cargo bikes with electric power assistance can be more cost-effective than lorries, in the case of small amounts, in densely populated urban areas [34]. In this paper, parking issues were highlighted as well, with cargo bikes having a huge advantage in this field over lorries.

As modeling is the primary focus of this paper, the most significant studies are those where the development and application of models for cargo bike-based city logistics systems are in focus. These models can have several different purposes, and they use several different objective functions. Some models can compare given scenarios, while in some others, they are trying to find the best solution to a problem by optimization. A study shows that the growth of the shares of CEP deliveries and home deliveries gives an extra reason to examine this field. In this paper, the situation of Munich was discussed, the optimal position of the centers was sought out with a view to optimal cargo bike routes, and the necessary delivery times and the effects of the new system were examined. In this paper, not only the mathematically optimal position for the depo was highlighted, but the greenest and most sustainable position, too. It was concluded that the sum distance could be reduced by integrating cargo bikes into the system, while the delivery times could still be met [35].

In a paper from 2011, the application of cargo bikes and micro consolidation centers in Manhattan (New York) was examined. In the investigated area, there are more than 100,000 delivery transactions each day, and based on their results, a significant part of them could be handled by electric cargo trikes—their integration for last mile deliveries would not cause an increase in costs, while the external effects could be reduced [36]. In a paper from 2017, the application of consolidation centers and cargo bikes in Sao Paulo was examined. In this city, TNT is currently operating cargo trikes for city logistics, while on most of the routes, there are problems with the use of lorries because of congestion, so the advantage of cargo bikes could be taken here, too [37].

Examining the case of London, the benefits of the application of cargo bikes are the reduction of vehicle purchase, delivery and parking costs, faster deliveries in congested city centers, and the reduction of emission. As a disadvantage, safety aspects and limited capacity were highlighted [32]. The expected benefits can be confirmed by the results of the pilot project from London in 2010. In this project, 7 fuel oil-powered lorries were replaced by 6 cargo trikes, 3 electric lorries, and 1 fuel oil-powered truck, adding a consolidation center to the system. In this project, in the urban areas, the mileage was reduced by 20%, and CO_2 emission was decreased by 54% [38].

In a study from 2019, the case of Seoul was examined, and similarly to our project, a city logistics system with cargo bikes and lorries was modeled. In this model, the share of cargo bikes and lorries was handled as an important strategic decision, and they analyzed scenarios by use of VRP algorithms (Vehicle Routing Problem) for the whole city. In their model, simulated annealing was used, and in the scenarios, more and more lorries were replaced by cargo bikes. Based on their results, the sum costs can be significantly reduced by adding cargo bikes; but it is also a significant result that not every lorry should be replaced by bikes because of their capacity. They concluded that CO_2 emission could be reduced approximately by 10% [39]. In a study from 2016, the combined application of cargo bikes and lorries was examined in Vienna. In this project, a two-stage model was developed, in which the different delivery vehicles are synchronized. They concluded, too, that emissions can be reduced by integrating cargo bikes. However, they say that the delivery costs can be increased [40]. A paper from 2018 focused mainly on restaurant deliveries by cargo bikes, examining them by simulation. In this research, lorries were added to the examined logistics system, which delivered the goods to urban consolidation points, and the last-mile delivery tasks were handled by cargo bikes. In this model, there is a new objective next to the minimization of the sum mileage: the minimization of the sum delays [41], which is very important in the case of delivering foods. The authors of this study published a simulation-based decision support system, too, and their results draw attention to the fact that, in these kinds of systems, it is essential to appropriate the number of available cargo bikes [42]. In another paper from 2017, a discrete-event-simulation-based (DES-based) solution was developed for the examination of the use of cargo bikes in multi-layer logistics networks, and the model was adapted for Grenoble. This model provides results about the daily time needs, demands, and the number of deliveries. Based on the results, it was concluded that CO_2 emissions could be reduced in this concept as well, and urban traffic jams will also be reduced [43].

The results of former research projects show that the city logistics use of cargo bikes can be various, and, from the modeling point of view, this area can be approached in many ways. Comparing the results of several projects, it can be concluded that in the case of the integration of cargo bikes, emissions and congestions could be reduced. However, mixed results can be seen for the different cost components: some of them can be reduced, but others are increasing in the new concepts. Based on the results of the literature review, it can be concluded that the geometrical model and the graph theory-based approach will be a completely new research direction, but several items from the results of previous research can be used, especially from models which examine lorries and cargo bikes together. It is also worthwhile to highlight that a graph-based approach to the cargo bikes-based city logistics systems would also be new, even though such an approach is used in transport and logistics research, for example, in the case of transport networks [44,45], inverse logistics aspects [46] or maintenance aspects [47].

Next, we would like to present the cargo bike-related results of our data collection, and we would like to summarize the results of the simulation project which motivated this paper.

2.3. *Possible Use of Cargo Bikes in the City Logistics System of Concentrated Sets of Delivery Locations in Budapest*

In the first phase of the research, 5 CSDLs in Budapest were examined, and we received data from about 540 stores by use of our own questionnaire [4], so the main logistics properties of 4 shopping malls and the properties of the stores of the biggest shopping area of the city are known. As in the case of the fourth shopping mall, not every cargo bike delivery-related question was included; only the data of 490 stores (327 stores from 3 shopping malls and 163 stores from a shopping area) are examined in this section.

First, we would like to highlight data about the delivery units. This question was answered by 475 stores of the 490. Of these, 417 (87.8% of the responders) are using (mostly small) boxes, and their quantity, size, weight, and volume are various—these are the most non-standard units in these systems. In one delivery transaction, an average of 16.6 boxes are arriving (standard deviation, 25.8), 5% of them are under 1 kg, 49.8% are between 1 and 5 kg, and 45.2% are more than 5 kg. This means a sum of 6 kg of goods in boxes every day per store (st. dev., 19.7 kg; yearly sum, 2.19 tons of goods in boxes/store). Most of these could be handled by cargo bikes, as the amount and their weight are mostly small. Of the responding stores, 12.8% are using pallet unit loads (there are stores that are using more delivery unit types); in one delivery, an average of 4.6 pallet unit loads are arriving (st. dev., 5.6). Of these, 42.9% are under 100 kg, and another 50% are under 500 kg, so most of these units could be handled by the three- or more-wheeled cargo bikes with a bigger capacity. Additionally, other, mostly small delivery units were mentioned by the stores, as can be seen in Figure 2.

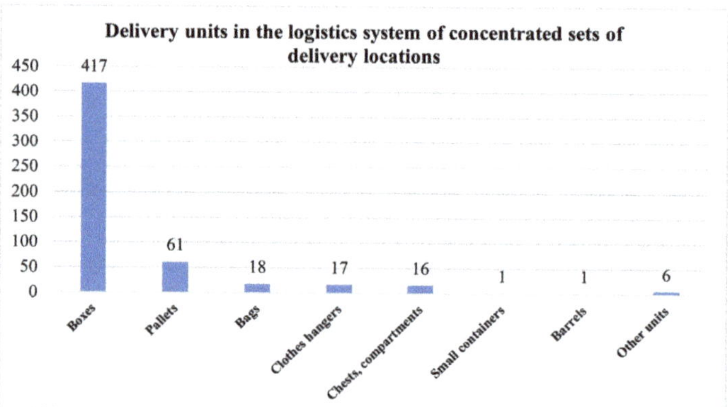

Figure 2. Delivery units in the logistics system of concentrated sets of delivery locations (N = 475).

The next important question was about the delivery vehicles currently being used. Most of the 474 responder stores are using lorries (277 stores, 58.4%) and their own passenger cars (171 stores, 36.1%) for the deliveries. These smaller vehicles could be easily replaced by the different types of cargo bikes in most cases, as most of the delivery units are small. Despite all of this, currently, only 5.1% of the stores (24 stores) are using bike or motorbike couriers for their deliveries.

It was examined what share of stores assumes that a given cargo bike (BULLITT two-wheeled bike, with a 50 × 50 × 70 cm delivery box) would be appropriate for their logistics transactions. Of the 375 responder stores, 29.3% answered that based on the size of their logistics units, the cargo bike could be used. Of these, 50% (55 stores) answered that the cargo bikes could be used for their incoming deliveries, and in the case of 20% (22 stores), they responded that the bikes could be used for home deliveries. A figure of 41.8% (46 stores) suggested that cargo bikes could serve the deliveries between their stores (see Figure 3). In the case of the incoming deliveries, an average of 3.91 daily deliveries by this cargo bike type would be necessary (st. dev., 5.29), and a maximum of 25 deliveries would be enough. In the case of 76.1%, a daily maximum of 3, and if 91.3%, a maximum of 10 daily deliveries would be enough. Based on their answers, 180 deliveries each day would be necessary to handle the demands of the 46 responding stores.

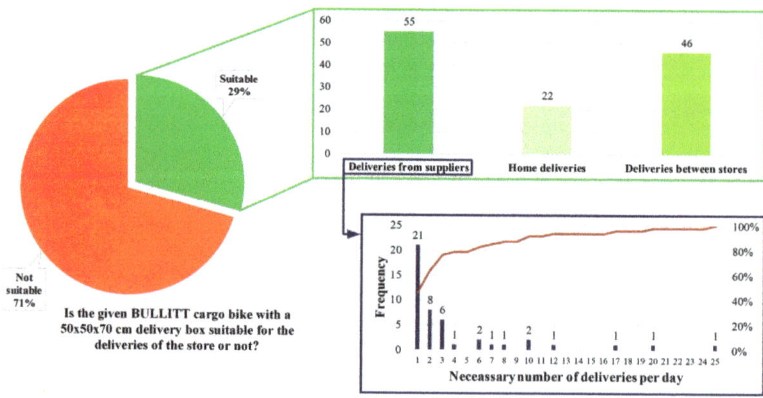

Figure 3. Suitability of a BULLITT cargo bike for different delivery tasks (N = 375).

It is essential to highlight here that this question was about one of the smallest available cargo bikes, so if the bikes with a bigger capacity were examined, the results would be even better, and a bigger suitability share could be seen. Based on these results, though, in the examined city logistics system, cargo bikes can be used for several different purposes. In this paper, we are going to discuss their integration into the delivery process of the CSDLs, in a well-defined, geometrical model-based city logistics concept. Next, we would like to present the simulation model from our former project and its results [9], which gave us the idea of this paper.

2.4. Application of Cargo Bikes in the City Logistics System of Shopping Malls in a Geometrical Model-Based Approach

Based on the previous section, it became clear that cargo bikes can become a part of the city logistics system of the CSDLs; only the deliveries are examined here, as they are the most significant logistics tasks. In a project in 2017, 18 shopping malls from Budapest were examined, together with a city logistics provider. In this project, it was expected that the existing logistics areas of the examined malls are appropriate to handle all the logistics tasks in the examined concepts, and it was also assumed that the shopping malls and the stores could work together in the new system. In this project, a geometrical model-based approach for city logistics modeling was developed. For this, two main groups were defined for the shopping malls (for the CSDLs, as the same concept can be suitable for any type, e.g., markets or shopping areas): there are malls on the central ring and malls on the rays that start from the central ring. In the simulation runs, it was examined how it is possible to integrate the deliveries of some stores of the shopping malls on the rays by use of cargo bikes (or use of electric lorries) to the deliveries of the malls on the central ring, in a gateway-concept based city logistics system. The primary purpose was to calculate the necessary number of vehicles in the examined concepts. The essence of the examined geometrical model-based solution is that instead of using VRP in the model, the geometrical model defines the locations, the connections, and the routes based on predefined rules, so the logistics network is given.

In this project, 4 different concepts were examined. In these concepts, the deliveries have been realized from the consolidation center next to the external ring of the city defined by the geometrical model, by the use of bigger electric trucks (EMOSS CM12 e-trucks in the simulation) to the central ring, so the central shopping malls have been served in this way. To these deliveries, the goods of those stores of the shopping malls will be added on the rays, in which case, the green small delivery vehicle on the rays (defined for each concept) is appropriate, based on the amount and size of goods. In the concept, these goods will be delivered from the central malls to the malls on the rays, using cargo bikes or electric lorries. The concept can be seen in Figure 4.

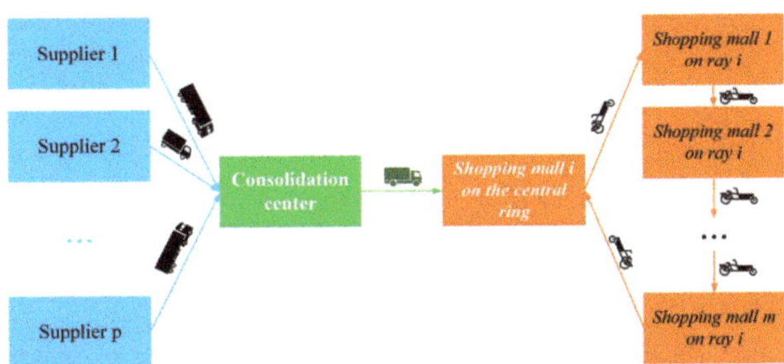

Figure 4. Examined new city logistics concept with cargo bikes.

In the first three concepts, we proposed the use of the previously examined BULLITT cargo bikes (with 50 × 50 × 70 cm delivery boxes) with electric power assistance, choosing those stores on the rays where a daily maximum of 1, 2, or 3 cargo bike deliveries are enough. In the fourth concept, the cargo bikes have been replaced by electric lorries (Nissan eNV-200 in the simulation model), choosing those stores on the rays where 1 delivery per day by lorry is enough. We developed a macroscopic simulation model to examine these concepts in MS Excel, which worked with aggregated data per shopping mall. The input data were provided by previous data collection; the data for every mall were estimated based on the data of 3 malls and their stores.

In this project, the primary purpose of the simulation runs was to receive data about the necessary number of vehicles in the examined concepts to handle the given delivery tasks. In every case, 12 e-trucks will be enough to handle all the delivery needs, so by adding more and more goods to the system, no more trucks are needed. Thus, with a relatively small investment cost, more and more stores could be served with the use of green vehicles: Additionally, 5, 9 or 13 BULLITT cargo bikes, or 8 electric lorries would be enough in the four concepts, and if cargo bikes are used, the load on the congested roads will become smaller [9].

As these simulation runs were performed for a given project, with given vehicle types, and with given shopping malls to serve, we decided to complete our model with all other malls from Budapest. It was also decided to use a new estimation methodology to provide input data for the model, and it was planned to examine several different types of cargo bikes in the new scenarios (with different sizes and capacities). Previously, only 18 shopping malls were examined and simulated [9], but in Budapest, there are 32 of this kind of CLSD, and most of them fit the previously examined rays or their extensions. All these malls and their geometrical model can be seen in Figure 5. The central ring is black on the figure, the rays start from it, grey means the city border, and blue is the external ring road of the city. Previously, data were collected about 4 of these 32 shopping malls, and in the case of 3 of them, data were collected about the cargo bike delivery-related questions as well.

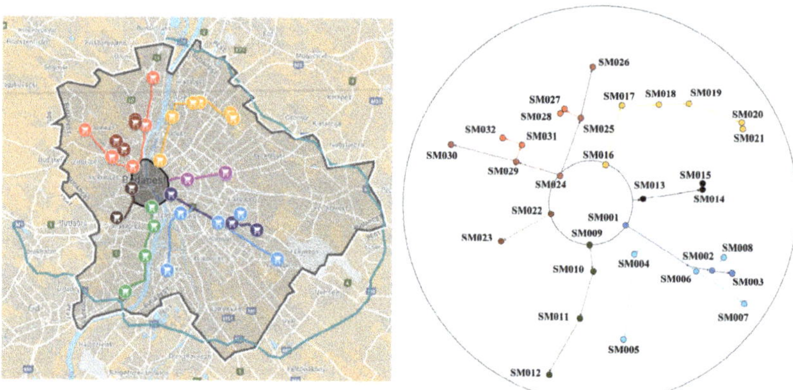

Figure 5. The examined shopping malls on Google Maps and the simplified version of the geometrical model.

For the estimation of the demands, an ABC-analysis-based [48] solution was used, where the malls were assigned based on the number of their stores to the categories "small", "medium", or "large". The number of deliveries and the demands was not considered for this solution (as there were no available data for every shopping mall), only the number of stores per shopping mall was examined, but it was assumed that the number of deliveries is proportional to the number of stores, as the types of the stores are similar in case of every shopping mall. The chosen limits were 33.33%, 66.67%, and 100%, so the stores of category A ("large") can give a maximum of one-third of all the stores of the malls from Budapest (and it was assumed that with this they give a maximum one-third of all the delivery demands of shopping malls from Budapest), and category A and B ("medium") can give a maximum two-third of the stores. The results of the analysis can be seen in Figure 6, where green is category A, yellow is category B, and red is category C. For anonymity reasons, the examined shopping malls are marked by their ID (identification numbers) in the paper.

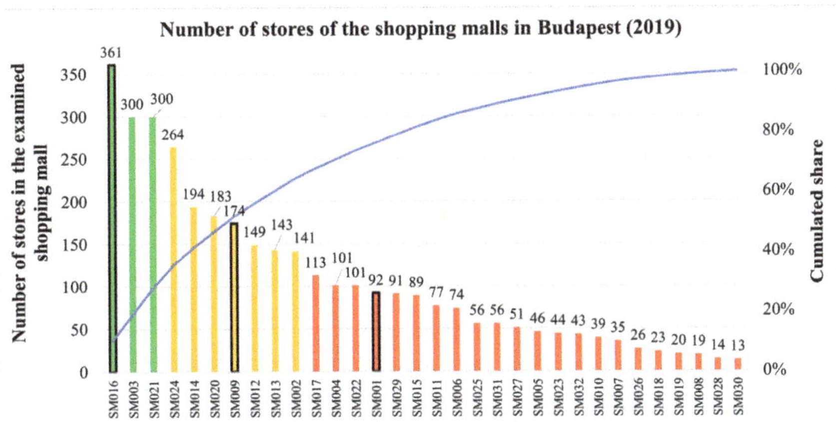

Figure 6. The number of stores and the ABC category of the examined shopping malls.

On the chart, the black outline shows those 3 shopping malls for which there were available data. As one of them is falling into every category, they can provide a reasonable basis for estimating the demand in the next steps. Based on the results of the former data collection, the expected characteristics can be seen below for each category:

- Category A: shopping malls with a big number of stores, mostly smaller stores with smaller demands. Their demands are expected to be served easily by cargo bikes.
- Category B: shopping malls with a middle number of stores, mostly stores with a bigger floor area and bigger demands. Their demands are expected more challenging to be handled by cargo bikes, but still, a significant share of the stores can be served by them.
- Category C: shopping malls with a low number of stores, most stores with a bigger floor area and bigger demands. Their demands are expected to be more challenging to be handled by cargo bikes, but there is still a share of the stores which could be served by them.

Based on this analysis, it was concluded that there is no correlation between the categories and the location of the malls in the geometrical model.

Next, the demands of all the malls were estimated based on the real data from its category, proportionally to the number of stores. First, specific values were calculated for all 3 categories based on the real data of the 3 shopping malls. The daily, weekly, monthly, and yearly expected number of deliveries were calculated, as well as the number of goods to be handled per day and per month, in kg and in m^3. We also examined with 4 cargo bike types the expected daily and the monthly number of deliveries, the number of stores to be served by them, and the expected daily and the monthly amount of goods to be delivered by them, in kg and in m^3. These 4 bike types will provide the 4 scenarios in the extended simulation model. To represent the main size-categories of the cargo bikes, 1–1 two-, three-, four- and six-wheeled solutions were chosen, and their parameters can be seen in Table 1. In the simulation, it was expected that the structure of each bike type is appropriate for the delivered logistics units, and the loading technology is available at every junction of the new system.

Table 1. The parameters of the examined cargo bike types.

Cargo Bike Type	STePS eBULLITT	Radkutsche Musketier Trike	Armadillo Cubicycle	Armadillo Cubicycle with Trailer
Wheels	2	3	4	6
Load capacity [kg]	100	242	125	250
Load capacity [m^3]	0.245	1.339	0.960	1.920
Number of delivery units	1	1	1	2
Length of the delivery unit—L [m]	0.70	1.27	1.20	1.20
Width of the delivery unit—B [m]	0.50	0.83	0.80	0.80
Hight of the delivery unit—H [m]	0.70	1.27	1.00	1.00
Battery [kWh]	0.504	0.540	1.200	1.200
Electric power assistance [W]	250	250	250	250
Average distance with one charge [km]	90	200	50	50
Charging time [h]	4	4	4	4

The estimated specific amounts for the 3 categories can be seen in Appendix A (Table A1), and the estimated amounts for the shopping malls (Tables A2 and A3) and the most important parameters (Table A4) can be seen in the tables of Appendix B. These values are macroscopic estimations, and the primary purpose of the estimation was to provide some data to map the real logistics processes, to examine the possible benefits of the examined concepts.

After estimating the demand values, simulation runs were performed by examining all the shopping malls from Budapest, with the geometrical model-based approach (see the geometrical model earlier in Figure 5; this geometrical structure is the one to be examined later). In the simulation model, the consolidation center was placed next to the external ring because of its good transport connections.

In this model, a total of 1135 stores of the 6 central shopping malls are served in each concept. Additionally, there are 26 shopping malls and 2297 stores within them on the rays. In the model, only those stores on the rays which could be served by the given cargo bike type were integrated into the new system (the other stores were not examined in these concepts; we examined only the central malls and those stores which can be added using cargo bikes to the new system). By use of eBULLITT cargo bikes, 342 stores can be served (14.9% of all); by use of Radkutsche trikes, 990 stores (43.1%); in the case of Armadillo cubicycles, this number is 805 (35%); and if a trailer is added to them to get a six-wheeled construction, 1078 stores can be served (46.9%, so nearly half of all stores on the rays). This already shows that by integrating cargo bikes into the system, more and more stores can be served without adding a significant number of extra trucks.

The simulation model was developed as an MS Excel-based macroscopic model, and the primary purpose was to get information about the necessary number of vehicles. A one-month-long period was modeled, and after the experiment design, we got the results that can be seen in Figure 7 for the 4 examined cargo bike types. In the simulation, the e-truck was again EMOSS CM12 type.

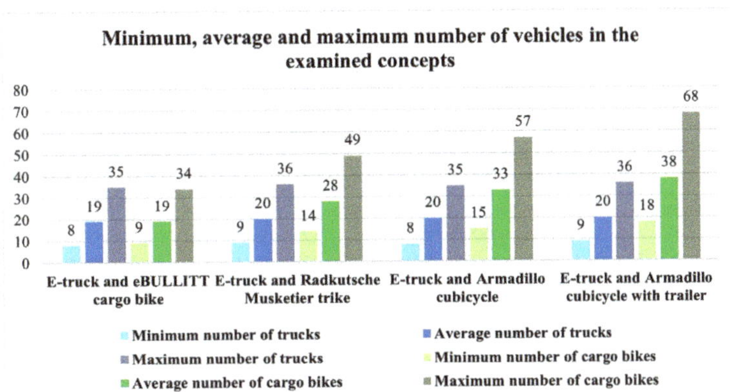

Figure 7. The number of vehicles, based on the simulation runs, for Budapest.

The necessary number of vehicles is not critically increasing; adding new stores in the distribution system has impact only on the number of cargo bikes significantly, similarly to the former results from 2017 [9]. If we would like to have a consolidation-based concept, where all the stores of each shopping mall are served by lorries (3432 stores), 60 trucks would be necessary (minimum 36, maximum 86) based on the simulation runs, but in the previously presented concepts, 19–20 trucks (minimum 8–9, maximum 35–36) would be enough, with additional cargo bikes. The necessary number of vehicles has a fluctuation in every case, as the demands are fluctuating in the model (as in the real system), but with organizational solutions, the necessary number of vehicles could be balanced around the expected value.

In the simulation, some other parameters were examined as well; they can be seen in Figure 8.

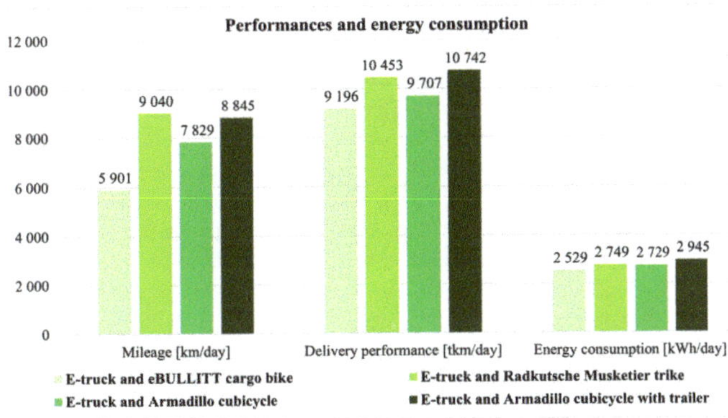

Figure 8. Mileage, delivery performance, and energy consumption based on the simulation runs.

Thus, we got positive results for the new system concepts. Based on the demands, a relatively large number of stores on the rays can be integrated into the system using cargo bikes, and the total number of vehicles will not be too large. It is also important that if cargo bikes are used instead of trucks and lorries, it also reduces the load of the overcrowded urban roads, as bike paths can be used instead of them, and the investment and operation costs of the bikes are also smaller. As there is one more stage in this solution than in the case of direct deliveries from consolidation centers to the stores, the delivery costs should be examined as well. It can be assumed that a geometrical model-based solution will not be cheaper, but there are several other factors that should be discussed. Considering the mathematical model of the cost structure of the concentrated sets of delivery locations [5], the delivery-related costs can be modeled. Based on the simulation runs, these costs were compared for each scenario with the costs of the direct deliveries from the consolidation centers to the concentrated sets of delivery locations by the previously mentioned electric trucks, in the case of the same amount of goods. In this macroscopic level model, the salary of the truck drivers, the salary of the bike couriers, the costs of the energy consumption based on the European Union average price from 2018 [49], and the other maintenance costs of the vehicles was examined. The results can be seen in Figure 9. In the figure, the column-pairs of the direct deliveries and the deliveries with cargo bikes are handling the same amount of goods; the bigger cargo the bike that is used in the system, the greater the amount that can be taken in the direct deliveries. In this comparison, it was assumed that the inventory-related, loading- and material handling-related and administration related costs would not be significantly different if the same amount of goods is handled at the same junctions (and the extra loading tasks in the central concentrated sets of delivery locations can be performed by the bike couriers).

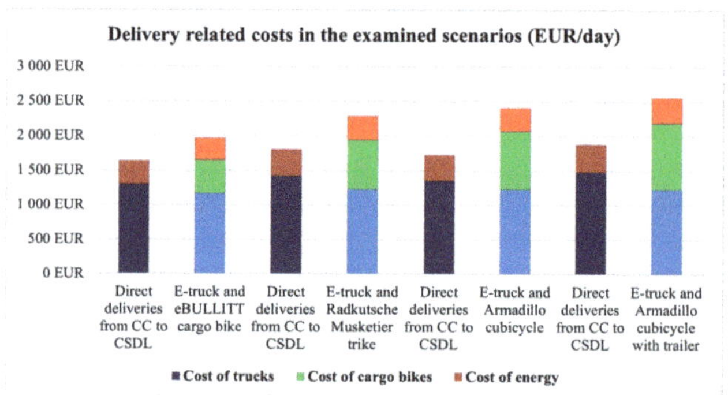

Figure 9. Comparison of the delivery-related costs between the examined scenarios and the direct consolidated deliveries.

The delivery-related costs will increase in the case of the solution with cargo bikes, as instead of direct deliveries, there will be an extra junction in the case of the CSDLs on the rays. The increase of the delivery-related costs is between 20% and 39.6% in the reviewed scenarios if the regular operation is examined, and it is more significant if more cargo bikes are needed. If only the financial reasons are considered, it is better to use the direct delivery solution, but if the potential effects of the traffic jams (late deliveries, extra deliveries needed with additional vehicles, etc.) are considered, the currently not significant difference can be even smaller. It must be highlighted as well that nowadays, it is getting more and more difficult to find suitably qualified drivers for lorries and trucks, and it is simpler to employ bike couriers. Especially if the development of the electric power assistance solutions will lead to faster deliveries, the number of required cargo bikes can be reduced, which can reduce the sum delivery costs as well.

So far, we have made estimations using a macroscopic simulation model to examine the effects of the geometrical model-based approach (and the new system concepts based on that). The model showed us how effective it could be to integrate cargo bikes into the examined city logistics system. The simulation provided lots of interesting data regarding this, and based on these results, it can be concluded that in such a city logistics system concept, a significant amount of goods could be handled, and a significant number of stores could be served by integrating cargo bikes, and the load of the urban roads could be reduced. It seems to be clear that the new solution is more expensive than the direct deliveries from the consolidation center to the concentrated sets of delivery locations, but the difference is less than 40% in every scenario, and it is also important that in the case of any traffic issues the cargo bikes can have better possibilities, and by use of the bike lanes, the urban roads will be less overcrowded. However, there is a significant number of additional questions related to these solutions: how to select the optimal central ring and the rays of the geometrical model, how to assign the CSDLs of the rays to the central ones, and when it is worth thinking about such a system at all. For the last question, we consider it essential to find a reference structure (optimal or close to it) that could be used as a benchmark in the future.

Next, we would like to examine the previously examined and described geometrical model based on graph theory, introducing its new notation. This will be the first step to find the reference geometrical structure, which can make it possible to evaluate cities or urban areas in the future and determine whether it might be worthwhile to examine such a concept in more detail for them or not.

3. Results: The Theoretical Geometrical Model and Its Application for Budapest

To describe the geometrical model based on graph theory, first, the relevant urban structures must be reviewed; next, we are going to define the theoretical, the symmetrical, and the real geometrical structure based on graph theory, and in the end, we are going apply the model and the notation to the shopping malls of Budapest as an example.

3.1. Urban Structures

To model the city logistics network, it is necessary to know the urban road network and the structure of the examined city. Examined from a morphological point of view, cities are usually composed of several different structures. For the current research, three main structural groups can be identified:

- irregular structure,
- grid-based structure, and
- radial structure.

In the case of an irregular urban structure, irregularity can mostly be found in the whole urban structure; the road network and the arrangement of the building are irregular as well. If the given city or given urban area has resulted from a conscious engineering design, there is a grid-based structure. In the case of radial urban structure, radially converging main roads and rings give the backbone of the urban structure [50], such as in the case of Budapest with "Kiskörút" and "Nagykörút" or in Vienna with the "Ring". The available road network provides a reasonable basis for the organization of the city logistics systems; the road network itself can also establish the logistics network, as the example of Budapest presented it previously. In the current case, the radial urban structure provides the basis of the geometrical modeling since the model that was examined earlier came from the radial structure of Budapest. In the previously examined model, we started from the benefits of the radial structure, taking advantage of the fact that the intersections of the rings and rays designate high-traffic nodes, and the rings and rays themselves define the main urban road network.

In general, there is a tendency for the CSDLs to be located in the main junctions (as they are transport junctions as well, this is ideal for customer traffic), and in the case of a radial structure, these junctions are provided by the intersections of rings and rays. The theoretical geometrical model, which is presented in the next sections, is based on the radial structure, but naturally, it can be adapted to any other structure type if a central ring and rays can also be designated on the road network, as can be seen in Figure 10. Later in the research, it will be an important task to develop the automatic, algorithm-based designation of the central ring and the rays (for example, by use of a genetic algorithm); in this paper, intuitive identification was used.

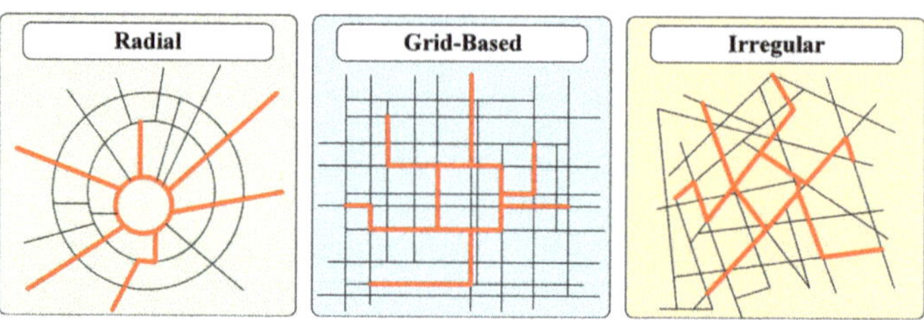

Figure 10. Designation of the radial structure-based geometric model in the case of different urban structures.

3.2. The Theoretical Geometrical Model

First, the theoretical geometrical model of the city logistics system of concentrated sets of delivery locations (in this case, primarily shopping malls) is defined. This theoretical model can be seen in Figure 11.

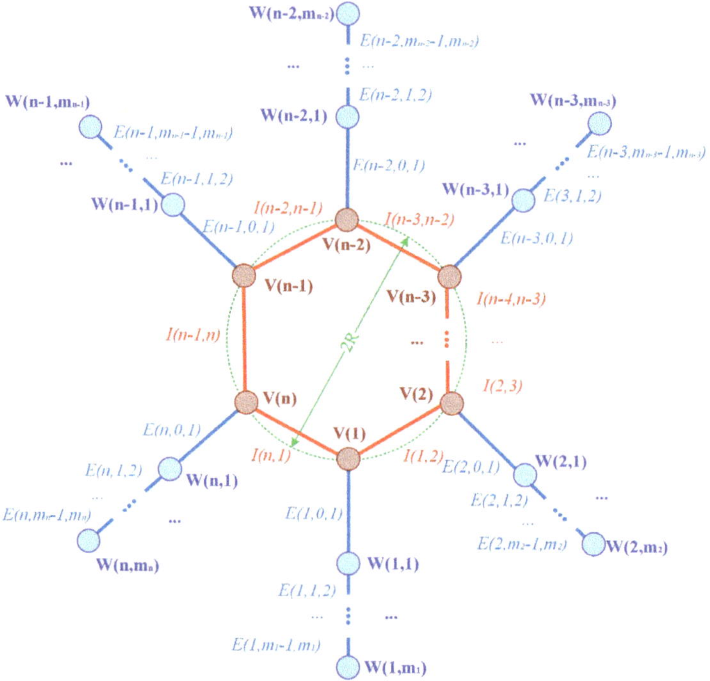

Figure 11. Theoretical geometrical model of the city logistics system of the concentrated sets of delivery locations.

The notation for the theoretical model can be seen below:

- n: number of vertices (CSDLs) on the central ring (given as polygons)
- m_i: number of vertices (CSDLs) on ray "i"
- $V(i)$: CSDL No. "i" on the central ring; its sum demand is $Q_{V(i)}$
- $W(i,j)$: CSDL No. "j" on ray "i"; its sum demand is $Q_{W(i,j)}$ (based on this logic, the CSDL on the central ring can be marked as $W(i,0)$)
- $I(i,i+1)$: the edge (road section) between $V(i)$ and $V(i+1)$ on the central ring; its length is $D[I(i,i+1)]$, $i = 1 \ldots n$; $I(n,n+1) = I(n,1)$

- $E(i,j-1,j)$: the edge (road section) between $W(i,j-1)$ and $W(i,j)$ on ray "i"; its length is $D[E(i,j-1,j)]$, $j = 1 \ldots m_i$
- R: the ray of the best fitting circle of the polygon of the central ring

In the theoretical model, the length of ray "i" can be calculated by the Formula (1):

$$d_i = \sum_{j=1}^{m_i} D[E(i,j-1,j)] \quad (1)$$

If the length of all rays is known, the average length of a ray in the examined structure can be calculated by the Formula (2):

$$D = \frac{\sum_{i=1}^{n} d_i}{n} \quad (2)$$

The average number of CSDLs on one average ray can be calculated by the Formula (3):

$$\overline{m} = \frac{\sum_{i=1}^{n} m_i}{n} \quad (3)$$

The average demand of an average CSDL on the central ring can be calculated by the Formula (4):

$$\overline{Q}_V = \frac{\sum_{i=1}^{n} Q_{V(i)}}{n} \quad (4)$$

The average demand of an average CSDL on ray "i" can be calculated by the Formula (5):

$$\overline{Q}_{W(i)} = \frac{\sum_{j=1}^{m_i} Q_{W(i,j)}}{m_i} \quad (5)$$

The average demand of an average CSDL of the examined geometrical model can be calculated by the Formula (6):

$$\overline{Q} = \frac{\sum_{i=1}^{n} Q_{V(i)} + \sum_{i=1}^{n} \sum_{j=1}^{m_i} Q_{W(i,j)}}{n + \sum_{i=1}^{n} m_i} \quad (6)$$

3.3. The Symmetrical Geometrical Model

Next, the symmetrical geometrical structure is defined, which can be seen in Figure 12, with the main parameters of the full symmetrical case.

The main properties of the symmetrical case can be seen below:

- the number of the CSDLs on the rays is the same on every ray (m)
- the length of the $I(i-1,i)$ edges is the same for every "i" value ($I = 1 \ldots n$), so the central ring will be a regular polygon
- the length of the $E(i,j-1,j)$ edge is the same on every ray, for every "i" and "j" value ($i = 1 \ldots n; j = 1 \ldots m$)
- the demands are the same for every CSDL, as in Formula (7):

$$\overline{Q} = Q_{V(i)} = \cdots = Q_{V(n)} = Q_{W(1,1)} = \cdots = Q_{W(i,j)} = \cdots = Q_{W(n,m)} \quad (7)$$

It can be assumed because of the symmetry that this symmetrical model will give later the reference geometrical structure that can be used as a benchmark. During the future comparison, it will be essential to get an answer as to what percentage of the current examined model approximates the reference structure, and based on this, it will be possible to decide on the suitability of the examined cargo bike-based concept for the given urban area. In the current case, the reference geometrical model will be a structure that is optimal in terms of well-defined aspects (e.g., sum mileage, the necessary number of vehicles, energy consumption, emission). It is important that in this case, we are looking for a reference model of given CSDLs and their relative position to each other, so a reference is needed that is adapted to the specifics of the examined system (e.g., in the case of a model with 6 central CSDLs and 20 CSDLs of the rays, we need a reference structure with 6 central CSDLs and 20 CSDLs on the rays). Later, in the next phases of the research, we would like to prove the reference structure in a heuristic way by use of the presented simulation model and in a mathematical way, too. It will be an essential question to find out if one "optimal" reference structure is there only or if it is possible to find alternative solutions as well.

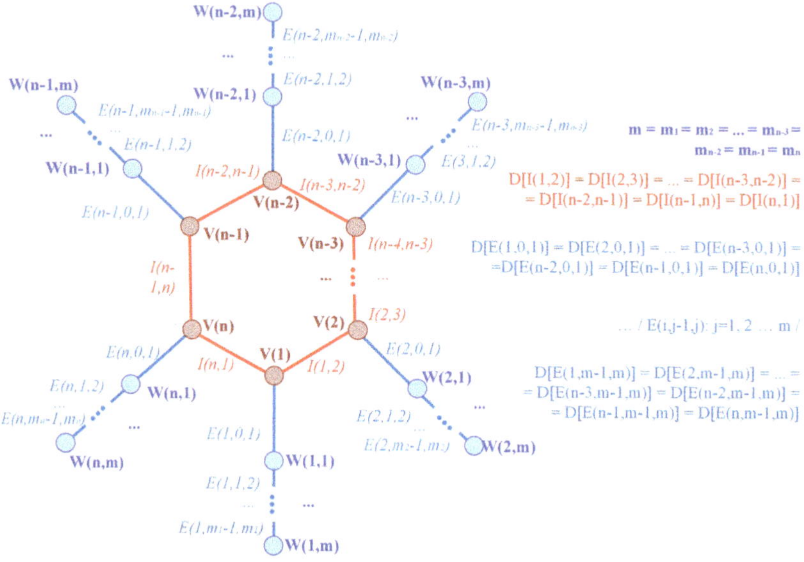

Figure 12. The symmetrical geometrical model of the city logistics system of the concentrated sets of delivery locations.

3.4. Geometrical Model of the Real Network

The real geometrical models are significantly different from the symmetrical model in every city logistics system; so, in real systems, the statements above are not true, the model is distorted, such as in the case of the geometrical model of the shopping malls in Budapest. In these real systems, there are also branches, i.e., certain sections of different rays are overlapping each other, and it is also possible to have central CSDLs, from which more than one ray starts. As can be seen in the example of Budapest, some sections of the rays are in the real system broken lines. The left side of Figure 13 (below) shows the model of a possible real system in the case of $n = 6$ (6 rays); the right side of the figure shows the simplified version of the model. It is important to highlight that because of the real structure of the transportation network, the rays and the central ring itself are given in the real model by broken lines, but they can be modeled as straight sections and as a polygon, with the original distances, as they are the critical indicators.

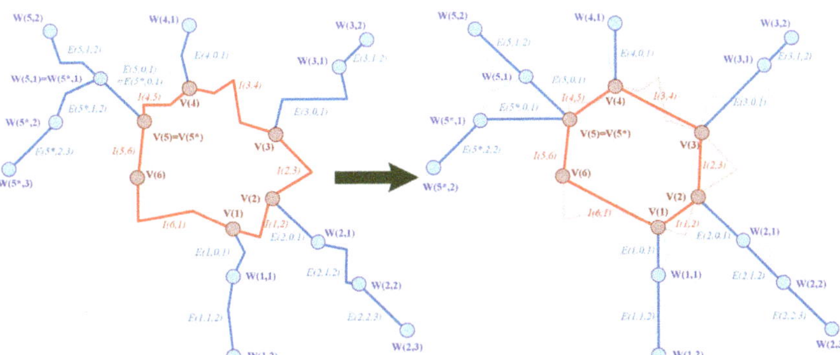

Figure 13. Geometrical model of a real city logistics system of the concentrated sets of delivery locations with slashes and then simplified, in the case of $n = 6$.

Thus, in these real cases, there can be such a central CSDL, from where more than one ray starts (these will be signed by the * character), and it is also possible to have CSDLs without further rays ($m_i = 0$). Thus, the number of rays will be more than, equal to, or less than the number of the central CSDLs (n). The general properties of such a real model can be seen below:

- the number of CSDLs on one ray is not equal in every case: in the example $m_1 = 2$, $m_2 = 3$, $m_3 = 2$, $m_4 = 1$, $m_5 = 2$, $m_{5^*} = 2$, $m_6 = 0$
- the length of the $I(i-1,i)$ edges is not the same in every case
- the length of the $E(i,j-1,j)$ edges is not the same in every case; there can be differences between the rays and on the rays as well

Next, we present the geometrical model with the new notation for the preciously examined example from Budapest.

3.5. Application of the Geometrical Model: Results for Budapest

After developing the graph theory-based notation of the geometric model, we are going to present it in a numerical example as well. The object of the numerical example is the previously examined network from Budapest. This is a good example of the real geometrical model, as the central ring and the rays can be clearly defined, but it is distorted at the same time—there are several branches in it, and one of the central CSDLs is not exactly fitting to the central ring (SM013). It should be mentioned that in the case of three pairs of shopping malls, the distance between them is close to zero, so they could even be served together with a common cross-dock in a new system. The examined geometrical model with the graph-theory-based notation can be seen in Figure 14. To the polygon of the central ring, an ellipse can be fitted instead of a circle, its major axis is 4.7 km, and its minor one is 3.5 km. On the figure, at the branches *, **, and *** are used to show those rays which belong to the same central CSDL.

For the examination of the geometrical model of the shopping malls in Budapest, all the main indicators were calculated (distances, goods amount, number of CSDLs) and were defined and described in the previous sections. The results can be seen in Table 2.

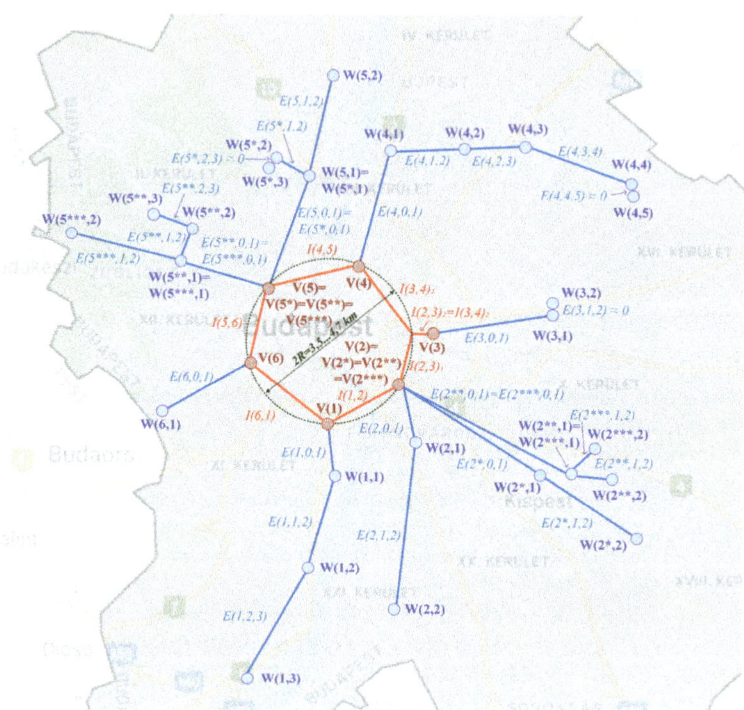

Figure 14. Geometrical model of the shopping malls in Budapest.

Table 2. Main parameters of the geometrical model of the shopping malls in Budapest.

Ray	Length of Ray i (d_i) [km]	Average Length of Rays (D) [km]	Number of Concentrated Sets of Delivery Locations (CSDLs) on Ray i (m_i)	Average Number of CSDLs per Ray ($\bar{m}$)	Number of CSDLs on the Central Ring (n)	Average Demand of a CSDL on the Central Ring ($\bar{Q}_V$) [t/month]	Average Demand of a CSDL on Ray i ($\bar{Q}_{W(i)}$) [t/month]	Average Demand of a CSDL in the System ($\bar{Q}$) [t/month]
1	8.84		3				926.4	
2	9.47		2				495.6	
2 *	11.44		2				367.5	
2 **	8.84		2				1248.9	
2 ***	7.54		2				1008.9	
3	3.80	7.46	2	2.42	6	1636.7	1600.1	841.3
4	11.10		5				822.5	
5	8.58		2				276.5	
5 *	5.76		3				272.0	
5 **	4.62		3				330.4	
5 ***	5.73		2				350.7	
6	3.85		1				296.7	

In Table 2, *, **, and *** are used to show those rays which belong to the same central CSDL. For example, Ray 5 is the first ray of central CSDL No. 5, 5* is its second ray, 5** is its third ray, and 5*** is its fourth ray.

The examined geometrical model is far from a symmetrical structure; additionally to the branches, the length of the rays, the number of CSDLs, and the estimated amount of goods are fluctuating. It was also examined whether any correlation could be detected between the length of the rays, the number of CSDLs on them, and the estimated average amount of goods, but the analysis showed no correlation; presumably, much more traffic characteristics, the size of the CSDLs, and the degree of concentration may be related to the quantities of goods.

Thus, we developed the theoretical, the reference, and the real geometric model of the city logistics system of the CSDLs and its notation based on graph theory, which was also applied to the examined Budapest example. Earlier, it was concluded, based on the simulation results using the specified geometrical model, that the cargo bike-based city logistics concept can be a right development direction because by integrating a relatively small number of cargo bikes into the system, many stores could be served. Additionally, a significant share of all freight demand in such a system could be served by using the bike path network instead of the already congested urban road network. By developing this complex model, and its exact notation, it became possible to examine the similar city logistics system in an exact way, and it became possible to examine that reference geometrical model in the future, which can later function as a benchmark in similar developments. Naturally, there are several further questions in this research which we would like to present next.

4. Discussion and Future Research

Previously, a simulation model was developed to examine a geometrical model-based new city logistics approach where cargo bikes were used in the logistics system; the theoretical model, the symmetrical model, and the model of the real system were developed, and this model was applied for the current network of the shopping malls of Budapest. Such an approach can be very useful in the future to plan new cargo bike-based city logistics systems and to integrate the use of cargo bikes into smart cities.

The primary purpose in the next steps is to define the reference geometrical model, which could be used as a benchmark in the future for the examination of any similar city logistics system. Knowing this reference geometrical structure, it will be possible to determine to what extent the reference structure approximates a given model, so it will be possible to see to what extent it is worthwhile to apply this geometric model-based city logistics approach for a given urban area. First, the plan is to define this structure by use of the presented simulation model in a heuristic way, and later we would like to prove its optimality in a mathematical way. There can be many aspects to define the reference structure; the proper selection of the objective function will be critical. It seems clear from the previous studies that multi-criteria examination will be necessary. Some ideas were already formulated (see Figure 15) for defining the reference geometrical model, along which the next research phase will be able to get started. The plan is to estimate the central ring with the best fitting circle/ellipse, so in the reference structure, the CSDLs will be moved to this circle; adding the rays to this, we are going to look after the smallest total squared deviation between the original and the new positions of the CSDLs, by rotation of the best fitting circle and the rays. In this model, the CSDLs can be moved to other rays if that would give the smallest deviation. Additionally, the optimal position of the consolidation center in the examined structure and the optimal number of consolidation centers will be important questions, too.

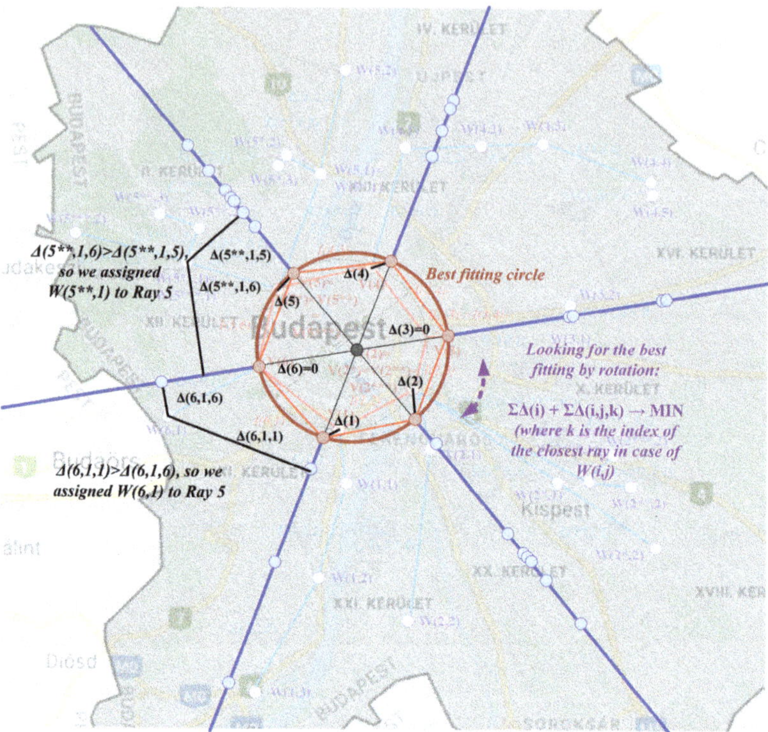

Figure 15. Designation of the reference geometrical model.

In this paper, primarily the case of Budapest and the theoretical model were examined, but in the future, the examination of other cities will be important too, and we would like to develop a ranking model which can help to decide, in the case of a given city, if it is worth implementing this geometrical model-based thinking or not. For this, the reference model will be able to help us. For some European cities, the examinations are already started, and based on them, Vienna and Warsaw seem to be good options for a similar concept. For this research, the analysis of the current logistics system and the estimation of the goods amounts will be very important for them, as based on our results, for example, in the case of Vienna, the characteristics of the shopping malls are much different to Budapest, and this must be considered at the estimation of delivery demands in the model.

It should be highlighted that in the case of the cities examined so far, there was no opportunity to evaluate the suitability precisely, so in the future, it will be very important to develop a ranking methodology to decide whether it is worthwhile to study such concepts in each city. Such a multicriterial method could significantly help in the future the application of this approach for a given city. Of course, in such a model, several indicators should be considered, and these indicators can have two main groups: geometrical model-based indicators and technology-based indicators.

In the case of the geometrical model-based indicators, the urban structure, the area to serve, the number of CSDLs, the properties of the model, and the network-related entropy-type indicators can be interesting. If the properties of the geometrical model will be examined, the length of the edges (the distance between the examined CSDLs on the central ring and on the rays), the covered central area, the number of CSDLs per rays, the number of rays, their length, and the number of CSDLs per rays should be highlighted. Here, it is also very important to consider the different characteristics of the different CSDLs of different cities, the different traffic demands, and their deviation as well. Later, if the currently two-dimensional geometrical model will be developed into a three-dimensional model, the terrain conditions (slopes) of the examined urban area will be important, it will be necessary to examine it in the ranking, with considering the altitude difference between the CSDLs on the central ring and on the rays.

The other group to examine is the group of technology-based indicators. In the ranking methodology, exact, modeled cargo bike types should be examined instead of real types. In general, the most important indicators are the capacity, interchangeability of the cargo boxes, and the range (with or without assistance). In the case of a cargo bike without electric power assistance, the human factor must be considered as well, and the examination of the meteorological conditions is important too, as in the case of worse weather conditions, the open structure of the vehicles can have several disadvantages. If the terrain conditions are going to be examined as well, the capability of these modeled cargo bike types to handle slopes must be examined. For this evaluation, the delivery technologies can have three main groups:

- two- or three-wheeled cargo bikes without electric power assistance;
- two-, three-, or four-wheeled cargo bikes (with or without trailer) with electric power assistance;
- non-pedal small electric delivery vehicles (e.g., cargo mopeds; they must be considered as their capabilities and advantages are very similar to cargo bikes)

In such a ranking methodology, several indicators must be examined, so complex multicriterial models are needed in the future. Primarily, the plan is to use the Analytic Hierarchy Process method, as it is appropriate to handle such complex problems. In the next steps of the research, the exact definition of the criteria and the formalization of the ranking model will be the primary tasks. Regarding our plans, this model will be able to show the suitability of given cities for a geometrical model-based city logistics approach with the use of cargo bikes by comparing them with their reference model.

In the future, it will also be essential to examine how to integrate the railway and waterway transport modes next to the road transport into the concepts of geometrical model-based thinking. In Figure 16 below, the previously examined system in Budapest has good railway and waterway transport connections, and the urban railways are even not marked—they can give lots of extra possibilities.

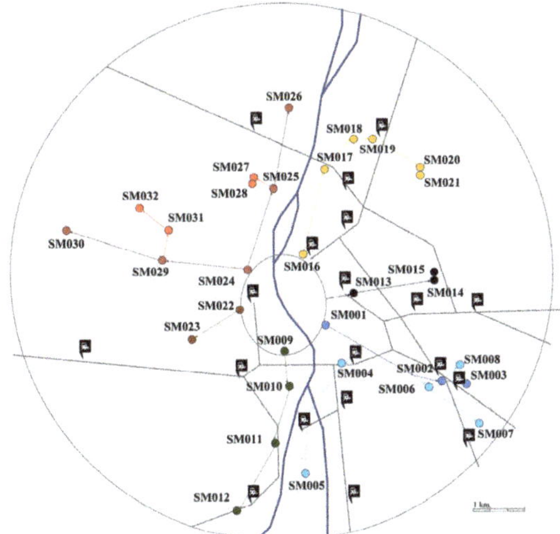

Figure 16. The geometrical model of Budapest, with railway and waterway connections.

In this paper, the city logistics system of the concentrated sets of delivery locations were examined, and only the business-to-business deliveries were included, from the suppliers to the stores of the CSDLs, via the consolidation center, using electric trucks and cargo bikes. Additionally, it can be very important to examine the other delivery types, primarily the business-to-customer deliveries (home deliveries, deliveries to parcel stations, or to parcel pick up points), but the deliveries between the stores of the same company and the deliveries to services should be examined as well. For these, the geometrical model and the simulation model must be developed; this can be a significant step for future research, as based on the analysis of the data from Budapest (3 shopping malls and 1 shopping area), 135 of 477 stores have business-to-customer deliveries (28%), 116 from 433 stores have incoming deliveries from other stores (27%), and 171 from 464 stores have outgoing deliveries to other stores (37%). If we would like to examine the business-to-customer deliveries, it will be important to investigate the so-called micro-hubs [51], as they can have a significant role in the geometrical model-based city logistics system with business-to-customer deliveries. Figure 17 shows those delivery transactions which must be added to the model in future research: the deliveries between the stores of the shopping malls by red lines and the home deliveries by blue lines; their geometrical model, graph-theory-based description, and the possible role of micro-hubs in the deliveries must be examined in the next steps.

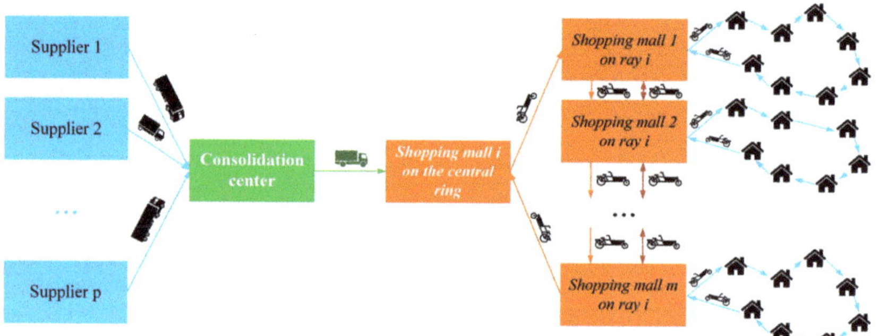

Figure 17. The examined concept with the deliveries between stores and home deliveries.

As a summary of the results of the paper, in Figure 18, the main steps of the application of the geometrical model-based new approach can be seen.

In this research, the development of the graph theory-based model and notation was motivated by the results provided by a previously developed simulation model. In future projects, the first steps will be to describe the geometrical model of the examined real system and to collect data about the CSDLs; based on these, it will be possible to develop a simulation model, and the results of the simulation together with the geometrical model of the real system and the reference geometrical model can provide data for the AHP-based multicriterial ranking which makes it possible to decide about the application of the new approach for given cities.

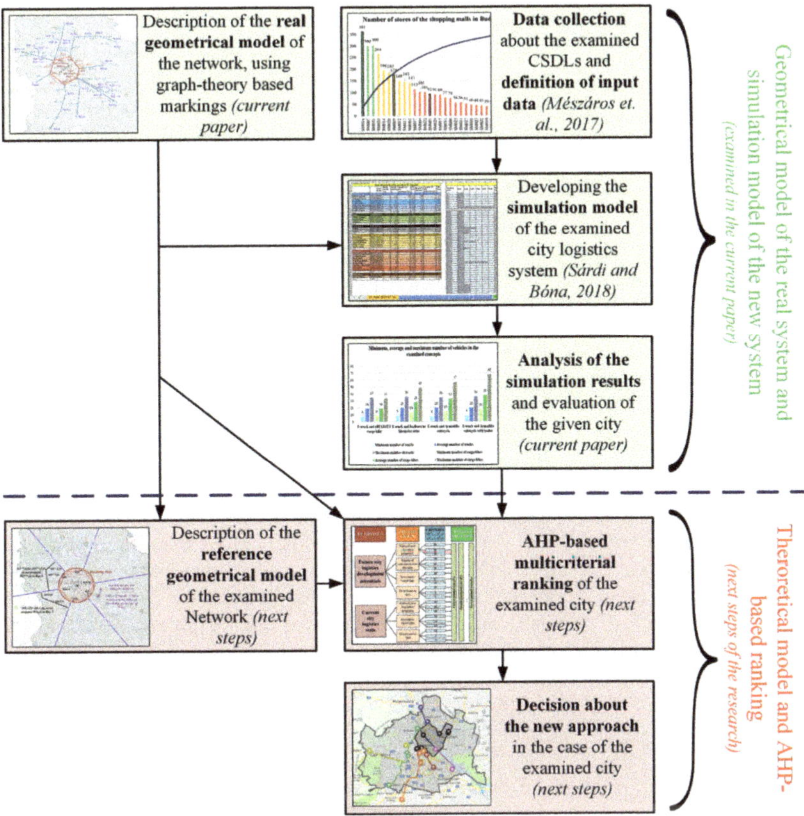

Figure 18. Application of the geometrical model-based approach.

5. Conclusions

In this paper, we examined geometrical model-based thinking in organizing city logistics by use of cargo bikes, in the logistics system of shopping malls, primarily for Budapest. This solution seems to be a new approach, based on the results of the literature review. Nowadays, these cargo bikes are more and more popular, especially in Europe, but in America and in Asia, they are playing an increasingly more significant role in city logistics. Such research is very important not only because of the current city logistics challenges but because of their possible integration to smart cities as well.

In the City Logistics Research Group of our Department [1], we are focusing on the so-called concentrated sets of delivery locations in our research project, and in this paper, we examined how cargo bikes can be used in their city logistics system based on the geometrical model of the logistics network. First, we presented the existing cargo bike-based systems and the related research results; next, our previous project was presented (from 2017). The results of this project gave the idea of this paper. In the concept, which was defined in this project, the CSDLs are divided into two groups: CSDLs on a central ring and CSDLs on rays, which start from the central ring, and we examined how to integrate the deliveries of the ones on the rays to the deliveries of the central ones.

For this research, the previously used simulation model was developed further and extended to the whole of Budapest (with all its shopping malls) by examining several different concepts using cargo bikes. Based on the results, it was considered that with a relatively small extra investment (so, with a relatively small number of additional cargo bikes), a significant amount of goods and a significant number of stores could be integrated into the new system, while the delivery costs will increase only 20–40%, and we can exclude the potential negative effects of the traffic jams on the rays (as they can cause longer delivery times, with more delivery costs). It is also important that it is simpler to hire bike couriers nowadays, and there are not enough available, suitably qualified drivers for trucks and lorries, so it can be possible that the logistics providers will be forced to use more cargo bikes; in this case, this geometrical model-based approach can help to organize their deliveries.

Next, the theoretical geometrical model, the symmetrical model, and the model of the real system were developed; with their graph theory-based notation, the description and application of such geometrical model-based solutions is an entirely new approach in city logistics planning. We applied this graph theory-based solution for the previously examined geometrical model in Budapest as a numerical example. In general, based on this geometrical model, it will be possible in the future to examine the suitability of given cities for implementation of the examined geometrical model-based new approach with cargo bikes.

In the last section of the paper, the planned next steps of the research project were presented. The first and most important task is to define the reference structure. We gave the related plans based on the example of Budapest. It is assumed that the symmetrical model can be the reference one, but alternative optimal solutions must be examined as well. Next, we would like to develop a ranking model, which will be able to rank given cities after comparing them with their reference models. This ranking model will be able to decide about the suitability of the examined cargo bike-based city logistics concepts for given cities.

The main result of this research project is the theoretical, symmetrical, and real geometrical model of the city logistics system of the concentrated sets of delivery locations, which provides an entirely new approach to the development of cargo bikes-based city logistics systems and gives the basis of our next research steps.

Author Contributions: Conceptualization, D.L.S. and K.B.; Data curation, D.L.S.; Formal analysis, D.L.S.; Investigation, D.L.S.; Methodology, D.L.S. and K.B.; Resources, D.L.S.; Software, D.L.S.; Supervision, K.B.; Validation, K.B.; Visualization, D.L.S.; Writing—original draft, D.L.S. and K.B. All authors have read and agreed to the published version of the manuscript.

Funding: This research received no external funding.

Institutional Review Board Statement: Not applicable.

Informed Consent Statement: Not applicable.

Data Availability Statement: Data is contained within the article and its appendices.

Conflicts of Interest: The authors declare no conflict of interest.

Appendix A. Specific Values for the Three Examined Categories

Table A1. Estimated specific values for the three examined categories.

Estimated Specific Parameters of the Categories		A	B	C
Number of deliveries/store	Daily	0.52	0.79	0.81
	Weekly	3.67	5.52	5.68
	Monthly	15.23	23.14	23.95
	Yearly	192.76	288.85	296.53
The daily amount of goods/store	Daily minimum [kg]	28.57	163.83	102.29
	Daily maximum [kg]	129.35	843.17	404.95
	Daily minimum [m^3]	1.20	3.59	2.00
	Daily maximum [m^3]	3.66	10.43	6.29
The monthly amount of goods/store	Monthly minimum [kg]	814.71	4851.19	3027.25
	Monthly maximum [kg]	3689.10	24,932.04	11,958.33
	Monthly minimum [m^3]	34.47	105.24	58.48
	Monthly maximum [m^3]	105.58	306.29	183.65
Necessary number of cargo bike deliveries/store	eBULLITT/day	0.11	0.11	0.04
	eBULLITT/month	3.22	3.02	1.03
	Trike/day	0.23	0.20	0.14
	Trike/month	6.43	5.72	3.90
	Armadillo/day	0.19	0.17	0.11
	Armadillo/month	5.42	4.96	3.04
	Armadillo with trailer/day	0.23	0.21	0.10
	Armadillo with trailer/month	6.51	6.06	2.73
Expected share of stores, they can be served by cargo bikes/store	eBULLITT	25.8%	20.3%	7.5%
	Trike	63.4%	55.4%	32.5%
	Armadillo	52.7%	45.9%	25.0%
	Armadillo with trailer	71.0%	63.5%	32.5%
Amount of goods to be delivered by cargo bikes/store	eBULLITT daily [kg]	1.09	0.26	0.05
	eBULLITT monthly [kg]	30.73	7.32	1.53
	Trike daily [kg]	11.23	5.83	4.44
	Trike monthly [kg]	321.10	167.30	124.88
	Armadillo daily [kg]	6.32	4.73	3.74
	Armadillo monthly [kg]	178.16	135.79	105.38
	Armadillo with trailer daily [kg]	13.70	8.86	4.44
	Armadillo with trailer monthly [kg]	391.69	253.41	124.88

Appendix B. Number of Stores, Deliveries and Amounts of Goods

Table A2. Number of stores, deliveries, and amounts of goods.

ID	Category	Number of Stores	Number of Deliveries			
			Daily	Weekly	Monthly	Yearly
SM001	C	92	68	470	1984	24,553
SM002	B	141	100	701	2936	36,656
SM003	A	300	142	992	4111	52,047
SM004	C	101	74	516	2178	26,955
SM005	C	46	34	235	992	12,277
SM006	C	74	54	378	1596	19,749
SM007	C	35	26	179	755	9341
SM008	C	19	14	98	410	5071
SM009	B	174	124	865	3624	45,235
SM010	C	39	29	200	841	10,409
SM011	C	77	57	394	1660	20,550
SM012	B	149	106	741	3103	38,735
SM013	B	143	102	711	2978	37,176
SM014	B	194	138	964	4040	50,434
SM015	C	89	65	455	1919	23,752
SM016	A	361	171	1194	4947	62,629
SM017	C	113	83	578	2436	30,157
SM018	C	23	17	118	496	6139
SM019	C	20	15	103	432	5338
SM020	B	183	130	910	3811	47,574
SM021	A	300	142	992	4111	52,047
SM022	C	101	74	516	2178	26,955
SM023	C	44	33	225	949	11,743
SM024	B	264	188	1312	5498	68,632
SM025	C	56	41	287	1208	14,945
SM026	C	26	19	133	561	6939
SM027	C	51	38	261	1100	13,611
SM028	C	14	11	72	302	3737
SM029	C	91	67	465	1962	24,286
SM030	C	13	10	67	281	3470
SM031	C	56	41	287	1208	14,945
SM032	C	43	32	220	927	11,476

Table A3. Estimated daily and the monthly amount of goods for the examined shopping malls.

ID	The Daily Amount of Goods				The Monthly Amount of Goods			
	Daily Min. [kg]	Daily Max. [kg]	Daily Min. [m^3]	Daily Max. [m^3]	Monthly Min. [kg]	Monthly Max. [kg]	Monthly Min. [m^3]	Monthly Max. [m^3]
SM001	8470	33,530	166	521	250,656	990,150	4843	15,207
SM002	20,791	106,999	456	1324	615,617	3,163,876	13,356	38,869
SM003	7715	34,926	324	990	219,971	996,057	9307	28,507
SM004	9298	36,810	182	572	275,177	1,087,012	5317	16,694
SM005	4235	16,765	83	261	125,328	495,075	2422	7604
SM006	6813	26,970	134	419	201,615	796,425	3895	12,231
SM007	3222	12,756	63	199	95,359	376,688	1843	5785
SM008	1750	6925	35	108	51,766	204,488	1001	3141
SM009	25,656	132,042	563	1633	759,697	3,904,358	16,482	47,965
SM010	3591	14,214	71	221	106,257	419,738	2053	6447
SM011	7089	28,063	139	436	209,789	828,712	4053	12,727
SM012	21,970	113,070	482	1399	650,545	3,343,387	14,114	41,074
SM013	21,086	108,517	463	1343	624,349	3,208,754	13,545	39,420
SM014	28,605	147,219	627	1821	847,019	4,353,134	18,376	53,479
SM015	8194	32,437	161	504	242,483	957,862	4685	14,711
SM016	9283	42,027	389	1191	264,698	1,198,588	11,199	34,303
SM017	10,403	41,184	204	640	307,871	1,216,162	5948	18,678
SM018	2118	8383	42	131	62,664	247,538	1211	3802
SM019	1842	7290	36	114	54,491	215,250	1053	3306
SM020	26,984	138,871	592	1718	798,992	4,106,307	17,334	50,446
SM021	7715	34,926	324	990	219,971	996,057	9307	28,507
SM022	9298	36,810	182	572	275,177	1,087,012	5317	16,694
SM023	4051	16,036	80	250	119,879	473,550	2316	7273
SM024	38,927	200,339	853	2478	1,152,644	5,923,853	25,007	72,775
SM025	5156	20,410	101	318	152,574	602,700	2948	9256
SM026	2394	9476	47	148	70,838	279,825	1369	4298
SM027	4695	18,588	92	289	138,951	548,888	2685	8430
SM028	1289	5103	26	80	38,144	150,675	737	2314
SM029	8378	33,166	164	516	247,932	979,387	4790	15,041
SM030	1197	4738	24	74	35,419	139,913	685	2149
SM031	5156	20,410	101	318	152,574	602,700	2948	9256
SM032	3959	15,672	78	244	117,155	462,788	2264	7108

Table A4. Suitability of cargo bikes, the sum, and the average amounts.

		Sum	Average (for One Shopping Mall)	Average (for One Store)
Number of stores		3432	107.3	1
Cargo bike deliveries	eBULLITT/day	258.3	8.1	0.1
	eBULLITT/month	7310.5	228.5	2.1
	Trike/day	570.2	17.8	0.2
	Trike/month	16,269.7	508.4	4.7
	Armadillo/day	477.3	14.9	0.1
	Armadillo/month	13,605.4	425.2	4.0
	Armadillo with trailer/day	540.3	16.9	0.2
	Armadillo with trailer/month	15,451.9	482.9	4.5
The average number of stores to be served by cargo bikes	eBULLITT	548	17.1	16.0%
	Trike	1545	48.3	45.0%
	Armadillo	1263	39.5	36.8%
	Armadillo with trailer	1699	53.1	49.5%
Amount of goods to be delivered by cargo bikes	eBULLITT daily [kg]	1290.1	40.3	0.4
	eBULLITT daily [m^3]	63.3	2.0	0.0
	eBULLITT monthly [kg]	36,477.1	1139.9	10.6
	eBULLITT monthly [m^3]	1790.9	56.0	0.5
	Trike daily [kg]	21,146.0	660.8	6.2
	Trike daily [m^3]	743.6	23.2	0.2
	Trike monthly [kg]	603,087.1	18,846.5	175.7
	Trike monthly [m^3]	21,214.5	663.0	6.2
	Armadillo daily [kg]	14,900.1	465.6	4.3
	Armadillo daily [m^3]	427.4	13.4	0.1
	Armadillo monthly [kg]	422,597.9	13,206.2	123.1
	Armadillo monthly [m^3]	12,185.2	380.8	3.6
	Armadillo with trailer daily [kg]	26,689.0	834.0	7.8
	Armadillo with trailer daily [m^3]	998.5	31.2	0.3
	Armadillo with trailer monthly [kg]	760,858.2	23,776.8	221.7
	Armadillo with trailer monthly [m^3]	28,549.1	892.2	8.3

References

1. BME-ALRT. Website of the City Logistics Research Group of the Department of Material Handling and Logistics Systems at the Budapest University of Technology and Economics. 2021. Available online: https://www.logisztika.bme.hu/citylog/ (accessed on 3 March 2021).
2. Sárdi, D.L.; Bóna, K. AHP-based multicriterial ranking model for the city logistics analysis of urban areas. In Proceedings of the 10th International Conference on Logistics, Informatics and Service Sciences (LISS2020), Beijing, China, 27–28 July 2020; Springer: Singapore, 2021. [CrossRef]
3. Sárdi, D.L.; Bóna, K. Examination of logistics systems of concentrated sets of urban delivery points by simulation. In Proceedings of the 21st International Conference on Harbor, Maritime and Multimodal Logistic Modeling & Simulation (HMS 2019); Cal-Tek Srl, Lisbon, Portugal, 18–20 September 2019. Available online: http://www.msc-les.org/proceedings/hms/2019/HMS2019.pdf (accessed on 3 March 2021).
4. Mészáros, B.; Sárdi, D.L.; Bóna, K. Monitoring, measurement and statistical analysis-based methodology for improving city logistics of shopping malls in Budapest. *World Rev. Intermodal Transp. Res.* **2017**, *6*, 352. [CrossRef]

5. Bóna, K.; Róka, Á.; Sardi, D.L. Mathematical Modelling of the Cost Structure of the Logistics System of Shopping Malls in Budapest. *Period. Polytech. Transp. Eng.* **2018**, *46*, 142–150. [CrossRef]
6. Yang, Z.Z.; Moodie, D.R. Locating urban logistics terminals and shopping centres in a Chinese city. *Int. J. Logist. Res. Appl.* **2011**, *14*, 165–177. [CrossRef]
7. Chiara, G.D.; Cheah, L. Data stories from urban loading bays. *Eur. Transp. Res. Rev.* **2017**, *9*, 50. [CrossRef]
8. Van Duin, J.; Van Dam, T.; Wiegmans, B.; Tavasszy, L. Understanding Financial Viability of Urban Consolidation Centres: Regent Street (London), Bristol/Bath & Nijmegen. *Transp. Res. Procedia* **2016**, *16*, 61–80. [CrossRef]
9. Sardi, D.L.; Bona, K. Macroscopic simulation model of a multi-stage, dynamic cargo bike-based logistics system in the supply of shopping malls in Budapest. In Proceedings of the 2018 Smart City Symposium Prague (SCSP), Prague, Czech Republic, 24–25 May 2018; Institute of Electrical and Electronics Engineers (IEEE): Piscataway, NJ, USA, 2018; pp. 1–7.
10. European Comission. White Paper 2011. 2011. Available online: https://ec.europa.eu/transport/themes/strategies/2011_white_paper_en (accessed on 3 March 2021).
11. Dong, X.; Wang, B.; Yip, H.L.; Chan, Q.N. CO_2 Emission of Electric and Gasoline Vehicles under Various Road Conditions for China, Japan, Europe and World Average—Prediction through Year 2040. *Appl. Sci.* **2019**, *9*, 2295. [CrossRef]
12. Lebeau, P.; Macharis, C.; Van Mierlo, J. Exploring the choice of battery electric vehicles in city logistics: A conjoint-based choice analysis. *Transp. Res. Part E Logist. Transp. Rev.* **2016**, *91*, 245–258. [CrossRef]
13. Csonka, B.; Havas, M.; Csiszár, C.; Földes, D. Operational Methods for Charging of Electric Vehicles. *Period. Polytech. Transp. Eng.* **2020**, *48*, 369–376. [CrossRef]
14. Thibbotuwawa, A.; Bocewicz, G.; Nielsen, P.; Banaszak, Z. Unmanned Aerial Vehicle Routing Problems: A Literature Review. *Appl. Sci.* **2020**, *10*, 4504. [CrossRef]
15. Kim, J.; Moon, H.; Jung, H. DroneBased Parcel Delivery Using the Rooftops of City Buildings: Model and Solution. *Appl. Sci.* **2020**, *10*, 4362. [CrossRef]
16. Kilián, Z. Környezetbarát Trendek a City Logisztikában. In Proceedings of the Nagyvállalatok Logisztikai Vezetőinek 9. Országos és 4. Nemzetközi Konferenciája, Eger, Hungary, 14 May 2015. Available online: https://docplayer.hu/2478203-Kornyezetbarat-trendek-a-city-logisztikaban.html (accessed on 3 March 2021).
17. Blaauw, E. DHL Lanceert Nieuwe Vervoerscombinatie Voor Stadsdistributie. 2017. Available online: https://www.dhlparcel.nl/nl/zakelijk/kennisplatform/nieuws/city-hub (accessed on 3 March 2021).
18. Chioda, E. Dall'Ungheria in Italia Inventa un Business con le Bici. 2015. Available online: https://www.millionaire.it/dallungheria-in-italia-inventa-un-business-con-le-bici/# (accessed on 3 March 2021).
19. Parr, T. RIPPL #31: UPS—Reducing Vehicle Movements with City Centre Container Hubs. 2017. Available online: https://www.rippl.bike/en/rippl-31-ups-reducing-vehicle-movements-with-city-centre-container-hubs/ (accessed on 3 March 2021).
20. Verlinde, S.; Macharis, C.; Milan, L.; Kin, B. Does a Mobile Depot Make Urban Deliveries Faster, More Sustainable and More Economically Viable: Results of a Pilot Test in Brussels. *Transp. Res. Procedia* **2014**, *4*, 361–373. [CrossRef]
21. Berlin.de. Modellprojekt in Berlin Startet: Lieferverkehr mit Lastenrädern Nachhaltig Gestalten. 2018. Available online: https://www.berlin.de/sen/uvk/presse/pressemitteilungen/2018/pressemitteilung.706285.php (accessed on 3 March 2021).
22. Bertram, I. Hermes Trials Parcel Delivery by Cargo Bikes in Berlin. 2018. Available online: https://newsroom.hermesworld.com/international/city-logistics-hermes-trials-parcel-delivery-by-cargo-bikes-in-berlin-837/ (accessed on 3 March 2021).
23. Vert Chez Cous. Our Vehicles. 2021. Available online: https://www.vertchezvous.com/en/nos-vehicules/ (accessed on 3 March 2021).
24. Janjevic, M.; Ndiaye, A.B. Inland Waterways Transport for City Logistics: A Review of Experiences and the Role of Local Public Authorities. *WIT Trans. Built Environ.* **2014**, *138*, 279–290. [CrossRef]
25. LogistikTram. Idee. 2021. Available online: http://www.logistiktram.de/index.html (accessed on 3 March 2021).
26. Giglio, C.; Musmanno, R.; Palmieri, R. Cycle Logistics Projects in Europe: Intertwining Bike-Related Success Factors and Region-Specific Public Policies with Economic Results. *Appl. Sci.* **2021**, *11*, 1578. [CrossRef]
27. Budapestcity.org. Kerékpár-közlekedés Budapesten—Kétkeréken a Városban. 2020. Available online: http://budapestcity.org/11-egyeb/kozlekedes/kerekpar/index-hu.htm (accessed on 15 July 2020).
28. Hajtás Pajtás. GLS—Hajtás Pajtás Összefogás a Kipufogógázmentes Csomagszállításért. 2015. Available online: https://hajtaspajtas.hu/blog/gls-hajtas-pajtas-oesszefogas-a-kipufogogazmentes-csomagszallitasert (accessed on 15 July 2020).
29. Schliwa, G.; Armitage, R.; Aziz, S.; Evans, J.; Rhoades, J. Sustainable City Logistics—Making Cargo Cycles Viable for Urban Freight Transport. *Res. Transp. Bus. Manag.* **2015**, *15*, 50–57. [CrossRef]
30. Cortés-Murcia, D.L.; Prodhon, C.; Afsar, H.M. The electric vehicle routing problem with time windows, partial recharges and satellite customers. *Transp. Res. Part E Logist. Transp. Rev.* **2019**, *130*, 184–206. [CrossRef]
31. Gruber, J.; Kihm, A.; Lenz, B. A new vehicle for urban freight? An ex-ante evaluation of electric cargo bikes in courier services. *Res. Transp. Bus. Manag.* **2014**, *11*, 53–62. [CrossRef]
32. Lenz, B.; Riehle, E. Bikes for Urban Freight? *Transp. Res. Rec. J. Transp. Res. Board* **2013**, *2379*, 39–45. [CrossRef]
33. Melo, S.; Baptista, P. Evaluating the impacts of using cargo cycles on urban logistics: Integrating traffic, environmental and operational boundaries. *Eur. Transp. Res. Rev.* **2017**, *9*, 30. [CrossRef]
34. Sheth, M.; Butrina, P.; Goodchild, A.; McCormack, E. Measuring delivery route cost trade-offs between electric-assist cargo bicycles and delivery trucks in dense urban areas. *Eur. Transp. Res. Rev.* **2019**, *11*, 11. [CrossRef]

35. Niels, T.; Hof, M.T.; Bogenberger, K. Design and Operation of an Urban Electric Courier Cargo Bike System. In Proceedings of the 2018 21st International Conference on Intelligent Transportation Systems (ITSC), Maui, HI, USA, 4–7 November 2018; Institute of Electrical and Electronics Engineers (IEEE): Piscataway, NJ, USA, 2018; pp. 2531–2537.
36. Conway, A. Urban micro-consolidation and last mile goods delivery by freight-tricycle in Manhattan: Opportunities and challenges. In Proceedings of the Transportation Research Board 91st Annual Meeting, Washington DC, USA, 22–26 January 2012. Available online: https://trid.trb.org/view/1129890 (accessed on 3 March 2021).
37. Telhada, J.; Duffrayer Ormond Junior, P.A. Using urban consolidation centre with cycloscargos support to increase the performance and sustainability of urban logistics: A literature review and a case study in the city of São Paulo. In Proceedings of the 3rd International Conference on Energy and Environment: Bringing together Engineering and Economics, Porto, Portugal, 29–30 June 2017.
38. Browne, M.; Allen, J.; Leonardi, J. Evaluating the use of an urban consolidation centre and electric vehicles in central London. *IATSS Res.* **2011**, *35*, 1–6. [CrossRef]
39. Lee, K.; Chae, J.; Kim, J. A Courier Service with Electric Bicycles in an Urban Area: The Case in Seoul. *Sustainability* **2019**, *11*, 1255. [CrossRef]
40. Anderluh, A.; Hemmelmayr, V.C.; Nolz, P.C. Synchronizing vans and cargo bikes in a city distribution network. *Central Eur. J. Oper. Res.* **2017**, *25*, 345–376. [CrossRef]
41. Fikar, C.; Gronalt, M. Agent-based simulation of restaurant deliveries facilitating cargo-bikes and urban consolidation. In Proceedings of the International Conference on Harbor Maritime and Multimodal Logistics, Budapest, Hungary, 17–19 September 2018.
42. Fikar, C.; Hirsch, P.; Gronalt, M. A decision support system to investigate dynamic last-mile distribution facilitating cargo-bikes. *Int. J. Logist. Res. Appl.* **2018**, *21*, 300–317. [CrossRef]
43. Hofmann, W.; Assmann, T.; Neghabadi, P.D.; Cung, V.D.; Tolujevs, J. A simulation tool to assess the integration of cargo bikes into an urban distribution system. In Proceedings of the 5th International Workshop on Simulation for Energy, Sustainable Development & Environment, Barcelona, Spain, 18–20 September 2017. Available online: https://hal.archives-ouvertes.fr/hal-01875988/document (accessed on 3 March 2021).
44. Miller, J. Vehicle-to-vehicle-to-infrastructure (V2V2I) intelligent transportation system architecture. In Proceedings of the 2008 IEEE Intelligent Vehicles Symposium, Eindhoven, The Netherlands, 4–6 June 2008; Institute of Electrical and Electronics Engineers (IEEE): Piscataway, NJ, USA, 2008; pp. 715–720.
45. Khan, Z.; Koubaa, A.; Farman, H. Smart Route: Internet-of-Vehicles (IoV)-Based Congestion Detection and Avoidance (IoV-Based CDA) Using Rerouting Planning. *Appl. Sci.* **2020**, *10*, 4541. [CrossRef]
46. Krikke, H.; van Harten, A.; Schuur, P. Business case Océ: Reverse logistic network re-design for copiers. *OR Spectr.* **1999**, *21*, 381–409. [CrossRef]
47. Čejka, J.; Kampf, R. The Winter Maintenance Optimization by Graph Theory. *Period. Polytech. Transp. Eng.* **2018**, *47*, 106–110. [CrossRef]
48. Flores, B.E.; Whybark, D.C. Multiple Criteria ABC Analysis. *Int. J. Oper. Prod. Manag.* **1986**, *6*, 38–46. [CrossRef]
49. Eurostat. Electricity Prices (Including Taxes) for Household Consumers, Second Half 2019. 2020. Available online: https://ec.europa.eu/eurostat/statistics-explained/index.php/Electricity_price_statistics#Electricity_prices_for_non-household_consumers (accessed on 3 March 2021).
50. Pirisi, G.; Trócsányi, A. Általános Társadalom- és Gazdaságföldrajz 3. Fejezet Településföldrajz—A Települések Szerkezete. Pécsi Tudományegyetem. 2018. Available online: http://tamop412a.ttk.pte.hu/files/foldrajz2/index.html (accessed on 3 March 2021).
51. Ballare, S.; Lin, J. Investigating the use of microhubs and crowdshipping for last mile delivery. *Transp. Res. Procedia* **2020**, *46*, 277–284. [CrossRef]

Article

A Multi-Criteria Decision-Making Approach for Ideal Business Location Identification

Salman Ahmed Shaikh [1,*], Mohsin Memon [2] and Kyoung-Sook Kim [1]

1 Artificial Intelligence Research Center (AIRC), National Institute of Advanced Industrial Science and Technology (AIST), Tokyo Waterfront, Tokyo 135-0064, Japan; ks.kim@aist.go.jp
2 Department of Software Engineering, Mehran University of Engineering and Technology, Jamshoro 76020, Pakistan; mohsin.memon@faculty.muet.edu.pk
* Correspondence: shaikh.salman@aist.go.jp; Tel.: +81-80-3557-4442

Abstract: Location has always been a primary concern for business startups to be successful. Therefore, much research has focused on the problem of identification of an ideal business site for a new business. The process of ideal business site selection is complex and depends on a number of criteria or factors. Since the ultimate goal of all businesses is to increase customer footprints and to thus increase sales, criteria including traffic accessibility, visibility, ease of access, vehicle parking, customers availability, etc. play important roles. In other words, we can say that optimal business site selection is a multi-criteria decision-making (MCDM) problem. MCDM is used to identify an optimal solution or decision out of many alternatives by utilizing a number of criteria. In mathematics, there exist a number of structured techniques for organizing and analyzing complex decisions, for instance, AHP, ANP, TOPSIS, etc. In this work, we present a hybrid of two such techniques to solve the MCDM problem for an optimal business site selection given a set of candidate sites. The proposed approach is based on the AHP (Analytic Hierarchy Process) and TOPSIS (The Technique for Order of Preference by Similarity to Ideal Solution) approaches. The reason for using the proposed hybrid approach is multi-fold. The hybrid approach reduces the computational complexity and require less manual effort, thus improving the efficiency and accuracy of the proposed approach. Given a set of candidate locations for a new business, the proposed approach ranks the candidates. Thus, the candidate locations with higher ranks are identified as suitable or ideal. The approach comes up with the ranking of all of the candidate locations, thus giving business managers room to make calculated decisions. To show the effectiveness of the proposed approach, a detailed step-by-step case study is given to identify an ideal location in New York City for a new gas station. Furthermore, an experimental evaluation is also presented using a number of real New York City datasets.

Keywords: multi-criteria decision-making; AHP/TOPSIS hybrid approach; optimal site selection; GeoSpatial data; smart cities

Citation: Shaikh, S.A.; Memon, M.; Kim, K.-S. A Multi-Criteria Decision-Making Approach for Ideal Business Location Identification. *Appl. Sci.* **2021**, *11*, 4983. https://doi.org/10.3390/app11114983

Academic Editors: Leon Rothkrantz, Miroslav Svitek and Ondrej Pribyl

Received: 8 April 2021
Accepted: 20 May 2021
Published: 28 May 2021

1. Introduction

The location of a brick and mortar business plays a vital role in its success or failure. In order to keep investors happy and to avoid any financial losses, it is necessary to select an optimal site for a new business. The term "optimal" refers to a location that may be suitable for a new business and that yields paybacks. However, the identification of optimal sites does not depend on any one factor. There are several aspects that require consideration, such as competition in the area, the target customers' convenience in terms of accessibility, the convenience of suppliers, traffic congestion in the area, etc. The selection of an optimal business site is a multi-criteria decision-making (MCDM) [1,2] problem. MCDM involves dealing with decisions where the choice of an alternative site is provided by several potential candidates while considering several criteria [3–5].

Since this problem poses great challenges, there are several related research papers that suggest different algorithms for the identification of an ideal location to open new business. Most of these approaches are either data based [6] or survey based [7,8]. Data-based approaches take into consideration only a single criteria in their evaluation, which results in biased decisions, whereas survey-based approaches lack the use of real data and are based on the opinion of a small group of people. Our argument is that, in big cities, for the selection of an optimal business site such as gas stations, convenience stores, restaurants, etc., a multi-criteria-based approach must be employed, and it should be applied on some real data for evaluation.

There is disagreement upon the decision of a suitable site for a new commercial opening as there are several criteria to be considered and some of them are more important/significant than others. For instance, for a gasoline station, ease of access for vehicles is an extremely important criteria, whereas for a convenience store, it is not so important. Based on the significance, different weights need to be assigned to different criteria. Hence, in this work, we present an AHP (Analytic Hierarchy Process) /TOPSIS (The Technique for Order of Preference by Similarity to Ideal Solution)-based approach [9,10], which choses an optimal site for a commercial opening when given a set of candidate sites. The techniques of AHP and TOPSIS are termed multi-criteria decision-making (MCDM) [1,2]. They are employed to derive criteria/factors' weight and to provide the ranking of alternatives/options. Although both approaches can be used to solve the MCMD problem separately, individually, each approach has some limitations. For instance, AHP alone is a very flexible and powerful MCDM tool and the computations made by AHP are always guided by the decision maker's experience. However, if the decision maker's understanding about the alternatives is not good enough, it can lead to inaccurate results. On the other hand, AHP alone requires a large number of evaluations by the user, especially for problems with many criteria and alternatives [11]. In fact, the number of pairwise comparisons grows quadratically with the number of criteria and options. This is discussed with an example in Section 4. To tackle this issue, Sangiorgio et al. proposed an optimized-AHP (O-AHP) method for the generation of a judgment matrix by using a mathematical programming formulation. According to the authors, O-AHP exhibits the same effectiveness as the standard AHP and can be easily applied to a large number of alternatives [12]. In [13], the authors use augmented reality-based decision-making (AR-DM) for the multicriteria analysis approach. The approach starts with problem structuring using a flowchart similar to that of the AHP and proceeds with new phases inspired from the SRF method. On the other hand, TOPSIS does not support criteria weight computation and must be combined with some technique to compute the weights. Usually AHP or entropy weights are used with TOPSIS [14]. Thus, in this work, we propose the use of an AHP/TOPSIS hybrid approach.

The hybrid approach results in (1) lowering of the computational complexity and (2) easing the manual effort needed for the construction of AHP pairwise comparison matrices. Here, all of the candidate sites are ranked based on their performance scores and the optimal site is identified as the candidate with the highest score. The decision makers are given flexibility to chose the optimal site in the form of top ranked alternatives based on the performance scores of all of the candidates/alternatives.

Since the AHP/TOPSIS approach consists of a number of steps, a detailed step-by-step case study is presented to identify an optimal location for a new gas station in New York City by employing four criteria. To assess the effectiveness of the proposed approach, an experimental evaluation is performed to select an optimal location of a convenience store making use of five criteria. The proposed approach presented here is flexible and can effectively incorporate any number of criteria to rank the candidates. This is an extended version of our work published in [15]. The main contributions of the extended work can be summarized as follows:

- A comprehensive flowchart showing the flow of important steps in the proposed AHP/TOPSIS approach (Section 5);
- An experimental evaluation utilizing real datasets to select an optimal location of a convenience store using the proposed AHP/TOPSIS approach (Section 6); and

- A detailed discussion on the evaluation results and the strength and weaknesses of the proposed AHP/TOPSIS approach (Section 7).

The rest of the paper is organized as follows: Section 2 discusses the related work. In Section 3, the problem formulation is presented. Section 4 elaborates two multi-criteria decision-making approaches. In Section 5, a case study is shown for the selection of an optimal location for a gas station. Section 6 presents experimental details for the evaluation of the proposed AHP/TOPSIS approach, while Section 7 discusses the evaluation results and the strength and weaknesses of the AHP/TOPSIS approach. Section 8 concludes the manuscript with the possible future directions.

2. Related Work

In this section, we present a few works related to optimal site selection of a commercial opening and the multi-criteria decision-making including AHP and TOPSIS approaches.

2.1. Computing the Best Alternative for a Commercial Opening

The research by J.Bean [16] provides an analysis of the prediction of a location for a new store by Starbucks. Various types of statistical analysis were applied for the identification of a suitable location to open a new Starbucks store. The results helped narrowed down the search for a location in the United States to find the best suitable location of a new Starbucks store. The authors were able to determine the success and popularity of existing businesses, difficulty accessing the store location, and peak rush hours. They also assisted in the development of a system to identify desirable locations for new businesses in that locality.

In [3], the authors presented a hierarchy of factors for selecting the best gas station site. In the study, the Analytic Hierarchy Process (AHP) methodology was used to calculate the relative importance of criteria and the sub-criteria in accordance with the aggregate opinions of experts. However, to compute the criteria in this work, the authors made use of a survey rather than real data as in our case.

The authors in [6] utilized machine learning features on the popularity of retail stores in the city through the use of a dataset collected from FourSquare. Their analysis is mainly focused on three different commercial chains, i.e., Starbucks, Dunkin Donuts, and McDonalds. The features that they mine are based on two general signals: geographic, where features are formulated according to the types and density of nearby places, and user mobility, which includes transitions between venues or the incoming flow of mobile users from distant areas. Their evaluation suggests that the success of a business may depend on multiple factors/criteria, which supports our study of a multi-criteria decision-making approach for an ideal business site selection.

The authors of [17] discussed a wide range of factors that are useful for decision-making such as whether it would be beneficial to open a commercial store in a certain locality. They identified the factors as being competition, vehicle ownership, and traffic rush in a locality, which may assist in making a decision about a new commercial store to ensure its success.

2.2. Multi-Criteria Decision-Making

In [18], the authors employed Fuzzy AHP for the selection of a new site for a hospital using the factors travel time and population density surrounding the new site of a hospital. The authors of [19] used AHP and spactial data in an attempt to determine candidate landfill sites. As a result, they were able to find the best, good, and unsuitable landfill areas. Awasthi et al. in [20] presented a TOPSIS-based approach for location planning under uncertainty. The uncertainty in their work was used to handle the lack of real data in location planning. In contrast, this work makes extensive use of real spatial data to compute alternative locations' factors/criteria computation. We believe that the use of real data can give us more accurate results. Furthermore, we made use of real spatial data to evaluate our approach.

Besides their individual use, the AHP/TOPSIS approach is frequently used together for different multi-criteria decision-making problems. For instance, the authors in [21] made use of the AHP/TOPSIS hybrid approach to identify and rank the solutions of RL (reverse logistics) adoption to overcome its barriers. Fuzzy AHP is applied to obtain the weights of the barriers as criteria by pairwise comparison, and the final ranking of the solutions of RL adoption is obtained using fuzzy TOPSIS. The authors in [22] utilized the AHP/TOPSIS approach to select the best alternative, with an aim to improve the electronic supply chain management (e-SCM) performance of an Indian automobile industry. Supraja et al. [23] utilized AHP/TOPSIS to solve the problem of selection of a branch of students for the "All Round Excellence Award" from an engineering college.

A CSPCM/TOPSIS approach for the quantification of accessibility to market facilities in rural areas was studied by Niaz et al. [24]. The authors made use of the Constant-Sum Paired-Comparison Method (CSPCM) to weight the factors and TOPSIS to rank the accessibility to market facilities. The study evaluated the accessibility of different urban and rural markets by using four factors, i.e., distance, time, cost, and road condition. Their study was mainly based on survey, i.e., a total of 335 questionnaire surveys were conducted from the whole study area (ten sub-districts) or, on average, 33.5 surveys per district, which is quite a small number for a district and is highly prone to bias. In contrast, we propose a real spatial data-based approach in this work, i.e., several million real dataset records and a spatial distance function are used to compute the criteria/factors weight. Furthermore, the accuracy of the obtained results are evaluated against real customer footprints.

Emrah et al. [25] proposed a supplier selection analysis model considering both the AHP and TOPSIS method. Subjective and objective opinions of purchase managers/experts are quantified using AHP. The TOPSIS technique is used for calculating the supplier's ratings. The aim of their research is to determine the appropriate supplier providing the most customer satisfaction for the criteria identified in the supply chain.

The proposed work presented in this paper provides a hybrid AHP/TOPSIS approach for pointing out an optimal site to open a commercial store. The reasons for selecting the AHP and TOPSIS methods are multi-fold. First and foremost is that the AHP and TOPSIS are among the most widely adopted MCDM techniques due to their simplicity and accuracy [26]. Secondly, when combined, they can reduce the computational efforts required to rank the alternatives. Other MCDM methods such as ANP, BWM, ELECTRE, or PROMETHEE would have been considered if one of our goals was not lowering the computational complexity and time required to find the ranking. Furthermore, TOPSIS works well with AHP quite nicely; that is, AHP is good for computing the criteria weights, while TOPSIS is good for ranking based on given weights. With other MCDM methods, such a hybrid approach is either not possible or not effective while non-hybrid approaches are computationally not feasible.

3. Problem Formulation

The proposed approach aims to identify the best location for a new commercial opening given a candidate set of locations. In accordance with this aim, the problem can be defined as follows:

Let us suppose that an enterprise wants to open a new business and is provided with a set of candidate locations L to chose from. The optimal location $l \in L$ such that the newly opened business at l attracts the largest number of customers is always preferred. Besides the identification of an optimal site, it is significant to rank each candidate site in L based on its performance score obtained using the AHP/TOPSIS hybrid method. The candidate locations ranking is generated by assigning top ranks to those locations that have higher scores. The location with the top rank is identified as the best site. Each candidate site in L is given in terms of the geographical coordinate system, i.e., in terms of longitude and latitude. Since AHP and TOPSIS are multi-criteria decision-making approaches and require multiple factors to compute the performance score of each candidate location, multiple spatiotemporal datasets are utilized to compute these factors.

4. AHP/TOPSIS-Based Multi-Criteria Decision-Making

The AHP (Analytic Hierarchy Process)/TOPSIS (Technique for Order of Preference by Similarity to Ideal Solution)-based hybrid approach is proposed in this research to compare a number of factors/criteria (in the following, criteria and factors are used interchangeably) and provides the rank based on those factors for each candidate. Many researchers have used either AHP or TOPSIS to find the best alternative from several potential candidates [3–5], since they are two different multi-criteria decision-making (MCDM) methods [1,2]. With the assistance of the approaches mentioned above, a set of criteria can be evaluated to rank the alternatives. The purpose of using AHP is to determine the criteria weights, and TOPSIS employs the weights obtained from AHP in order to rank the alternatives; however, on many occasions, the two approaches are combined to obtain optimal results [27]. AHP/TOPSIS refers to the mechanism where TOPSIS uses AHP weights. The reasons behind the use of this hybrid approach in this article, are as follows:

- It helps lower the computational complexity.
- It eases the manual effort needed for the construction of the AHP pairwise comparison matrices.

Although AHP alone is a very flexible and powerful MCDM tool, the computations made by AHP are always guided by the decision maker's experience, and AHP can thus be considered a tool that is able to translate the evaluations made by the decision maker into a multi-criteria ranking. However, if the decision maker's understanding about the alternatives is not good enough, it can lead to inaccurate results. On the other hand, AHP alone requires a large number of evaluations by the user, especially for problems with many criteria and alternatives [11]. In fact, the number of pairwise comparisons grows quadratically with the number of criteria and options. Let us discuss the following example to understand the proposed approach.

Example 1. *In this example, the best alternative is computed from a set of 10 alternatives using five criteria. Only two alternatives are taken at a time by AHP and a comparison matrix is constructed by comparing their criteria one by one. As a result, the best one is found among the 10 alternatives via AHP. A total of* $\binom{10}{2} = 55$ *combinations are generated, i.e., a total of 55 comparison matrices need to be constructed. Each criteria is given some weight based on its importance. Hence, a comparison matrix of the criteria is necessary to obtain criteria weights. In total, 56 comparison matrices are required and each matrix has to be made with careful selection of the degree of importance for each matrix cell [11]. This process takes a lot of time, and the problem intensifies when the number of alternatives are increased. Once TOPSIS is applied after given the criteria weights, a single decision matrix is generated with the assistance of all of the alternatives simultaneously. The best alternative or ranking of the alternatives is generated with the help of this decision matrix.*

The TOPSIS approach drastically reduces the computational complexity and the manual effort required in the AHP approach, but it does not assist in the formulation of a mechanism to compute the criteria weights.

Several researchers have adopted hybrid MCDM approaches to increase the overall efficiency. In [28], the authors suggested that the hybrid between AHP and Complex Proportional Assessment (COPRAS) to rank seven sustainable hydrogen production options helped reduce uncertainties in all phases of the task. In another research [13], AHP and SRF were combined to determine precast concrete panels for building retrofitting. They defined six criteria and evaluated seven visual rankings within 15 min, making it easier even for non-expert users to carry out the procedures. Sedghiyan et al. in [29] ranked seven alternatives of renewable energy sources with AHP, TOPSIS, and Simple Additive Weighting (SAW). The best renewable energy source in varied climate zones was identified if it was ranked best by any two methods. In [30], ten high-risk activities in the mining sector were identified in order to access them in a work environment. The risks were associated with multiple activities to generate a house of safety with the assistance of AHP and fuzzy inference. Keskin et al. in [31] proposed a hybrid AHP and Data Envelopment Analysis-Assurance Region (DEA-AR) model to measure the efficiency of public and private airports in Turkey. A total of five criteria were chosen, and 48 airports were ranked based on these criteria. In another research [32], nine risk criteria were chosen and five risk response alternatives were ranked in construction projects. They employed the Analytical Network Process (ANP) and Multi-Attributive Border Approximation Area Comparison (MABAC) techniques to reduce imprecision and fuzziness in the decision process. The work of [33] used a hybrid of AHP and Genetic Algorithms (GAs) to access ground water vulnerability in China. They optimized the land ratings from 1 to 10, where 10 indicates the highest potential to pollution, based on eight factors. In [34], Fuzzy AHP and Particle Swarm Optimization (PSO) techniques were used to solve the optimization model as a nonlinear system of equations. Their proposed method was scalable and was applied on various cases studies for prioritization even with an incomplete set of judgments. Table 1 compares the various hybrid MCDM approaches along with their applications, strengths, and weaknesses.

Table 1. Hybrid MCDM approaches and their applications, strengths, and weaknesses.

Ref.	Hybrid MCDM Approach	Application(s)	Strength(s)	Weakness(es)
[28]	AHP and Complex Proportional Assessment (COPRAS)	Sustainable hydrogen production options	(1) Controls uncertainties in all phases. (2) Improves group decision-making.	Relied on data given by only four experts of the domain.
[13]	Augmented Reality (AR) and the Simos-Roy-Figueira (SRF)	Building retrofitting	(1) Visual information during the decision phase. (2) On average, takes 15 min to evaluate seven rankings.	The technique cannot be applied to non-graphical problems where 3D models are not required.
[29]	AHP, TOPSIS, and Simple Additive Weighting (SAW)	Renewable energy resources prioritization	(1) The best alternative is the one that is selected best by at least two methods. (2) Consumes a lot of time since ranking is generated using three methods.	
[30]	AHP, House of Safety, and Fuzzy logic	Risk Assessment in Mining Sector	(1) Dependency of multiple events is considered. (2) Imprecise and vague structures are managed.	Results highly reliant on the input of occupational safety specialists.
[31]	AHP and Data Envelopment Analysis-Assurance Region (DEA-AR)	Efficiency scores of publicly or privately operated airports	Prevents extreme weight distribution.	The model can be discriminatory in some situations.
[32]	Analytical Network Process (ANP) and Multi-Attributive Border Approximation Area Comparison (MABAC)	Uncertain risk strategies in construction projects	Addresses uncertain information weighting risk criteria such as incompleteness and imprecision.	Computational time is high.
[33]	AHP and Genetic Algorithms (GAs)	Groundwater vulnerability assessment	Better results compared to DRASTIC [35] model.	Other contaminants except nitrate are not considered in this study.
[34]	Fuzzy AHP and Particle Swarm Optimization (PSO)	Professor selection/Investment prioritization	(1) Applied to group decision-making. (2) Solves the optimization model as a nonlinear system of equations.	PSO has a tendency to converge prematurely on early best solutions.

This work proposes the use of an AHP/TOPSIS hybrid approach. The AHP/TOPSIS hybrid approach efficiently caters to weight computation and the problem of AHP having a large number of matrices to compute, i.e., using the hybrid approach, the criteria weights are determined by AHP while the alternatives are evaluated using TOPSIS. In evaluating the alternatives, the TOPSIS approach makes use of the weights computed via the AHP approach. Thus, for instance, if there are 10 alternatives, then the AHP approach alone requires 56 comparison matrices. Using the AHP/TOPSIS approach, this is reduced to two matrices, i.e., one matrix for criteria weight computation using AHP and the other matrix for alternative/candidate evaluation using TOPSIS. A short description of the two approaches is given in the following subsections, while the detailed step-by-step procedure of the proposed hybrid approach for the selection of an optimal site for a new business is presented with the help of a case study in Section 5.

4.1. *Analytic Hierarchy Process (AHP)*

The organization and analysis of complex decisions of mathematics and psychology is best achieved via the analytic hierarchy process (AHP). It was developed by Thomas L. Saaty [9,11] in 1970s and has been extensively studied and refined since then.

A set of evaluation criteria and alternative options are considered by the AHP, among which the best has to be selected. A weight for each evaluation criterion is generated by the AHP according to the decision maker's pairwise comparisons of the criteria. The higher the weight, the more important the corresponding criterion is. In the next step, for a fixed criterion, AHP assigns a score to each option according to the decision maker's pairwise comparisons of the options based on that criterion. The performance of the option is directly proportional to the score with respect to the considered criterion. Finally, the criteria weights and the options scores are combined by the AHP, which helps in assigning a global score to each option and eventually a consequent rannking. The global score for a given option is a weighted sum of the scores obtained with respect to all of the criteria [11]. The following are the steps taken during the AHP process.

1. The problem is defined, the different criteria that are essential to decision-making are selected, and a hierarchical model is developed.
2. A set of pairwise comparison matrix is constituted with the assistance of the criteria, and priorities (weights) are derived.
3. The consistency of the comparison matrix is checked.
4. Final decision is made based on the derived priorities.

4.2. *The Technique for Order of Preference by Similarity to Ideal Solution (TOPSIS)*

One of the multi-criteria decision-making (MCDM) methods is the Technique for Order of Preference by Similarity to Ideal Solution (TOPSIS), which was created in 1981 by Ching-Lai Hwang et al. [10]. It is a compensatory aggregation method based on the idea that the ideal candidate or alternative must have the smallest geometric distance to a PIS (positive ideal solution) and the geometric farthest distance from an NIS (negative ideal solution) [36]. In other words, the benefit is maximized and cost is minimized by PIS; on the other hand, the benefit is minimized and cost is maximized by NIS. It is assumed that, for each criteria, maximization or minization is applied. TOPSIS requires normalization as the parameters or criteria are often of incongruous dimensions in multi-criteria problems [10,37]. Ranking a number of feasible alternatives based on the closeness to the ideal solution is best achieved via the TOPSIS approach. TOPSIS also avoids pairwise comparisons, thus allowing it to be computed in a simple and efficient manner [25]. The TOPSIS method implementation can be summarized with the following steps [38]:

1. Establish a decision matrix;
2. Calculate a normalized decision matrix;
3. Determine the weighted decision matrix;
4. Identify the positive and negative ideal solutions;
5. Calculate the separation distance of each competitive alternative from the ideal and non-ideal solutions;
6. Measure the relative closeness of each location to the ideal solution; and

7. Rank the preference order.

5. Case Study: An AHP/TOPSIS-Based Optimal Gas Station Site Selection

This section presents a case study to identify an optimal site for a new gas station. The case study makes use of the AHP/TOPSIS hybrid approach to compute the criteria weights and to choose the best option from among a finite set of decision alternatives. The four criterion values are computed using four real geospatial datasets. Although there could be several criteria affecting the new business site selection for an optimal gas station site, we followed the results of [7]. The authors presented a multi-criteria factor evaluation model for a gas station site selection. Their study was mainly based on a questionnaire/survey and they identified several important and crucial factors for site selection. Figure 1 presents a flowchart depicting the flow of important steps in the proposed AHP/TOPSIS approach. Among the factors they identified, we adopted the following four important factors in this study, listed with respect to their significance from top to down:

- Competitors in the area (competition);
- Traffic near the location (traffic);
- Popularity of the close-by businesses (area-popularity); and
- Approximate number of vehicles nearby residents own (vehicle-owners).

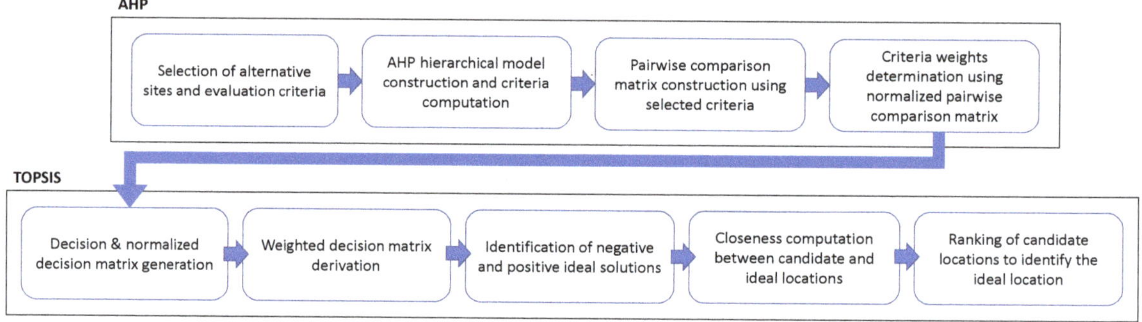

Figure 1. Flow of the important steps in the AHP/TOPSIS approach.

The appraoch in this study is empirical in contrast to the appraoch by [7] for the identification of a feasible location for a new gas station. In the Section 5.1, all of the datasets used in the study along with the prediction criteria are listed.

5.1. *Dataset Analysis and Prediction Criteria*

Four different geographical datasets used in this work are presented here for the computation of four different criteria.

5.1.1. Gas stations Dataset and the Competition Criterion

The NYC Open Data [39] provides the New York City gas stations data. Detailed addresses of the 416 gas stations present in New York City are provided in this dataset and shown in Figure 2. In this study, the competitors that are near the candidate gas stations are computed with the help of the reference dataset. According to [7], one of the most significant factors in identifying the ideal location for opening a gas station is the competitors in that area. The best location for a new gas station is the location with the least number of competitors.

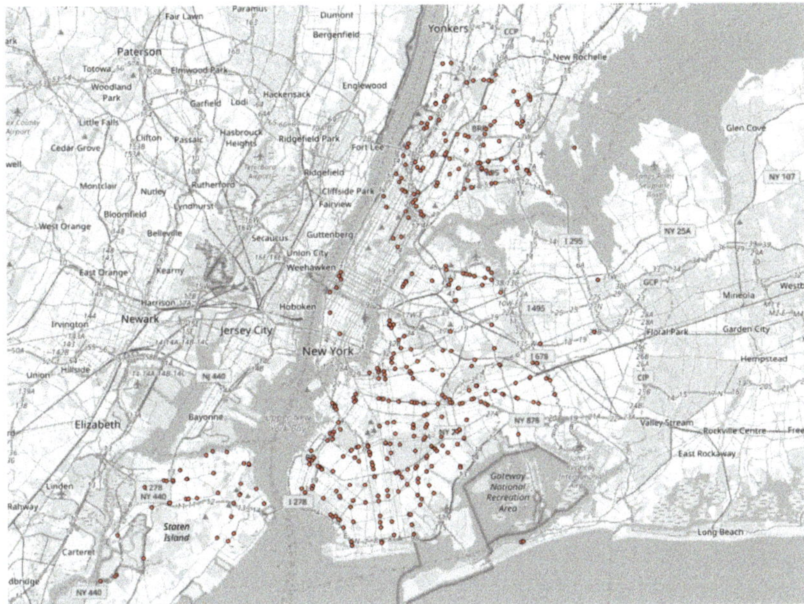

Figure 2. NYC gas stations.

Let us suppose that G represents the set of all gas stations in New York City, while g refers to the geographical location of a gas station. The competition criterion can be computed with the help of the number of gas stations within radius r_g of each candidate location $l \in L$. Equation (1) explains the abovementioned appraoch.

$$|\{g \in G : dist(g,l) < r_g\}| \quad (1)$$

where the Euclidean distance between locations a and b is represented by $dist(a,b)$.

5.1.2. Traffic Estimates Dataset and the Traffic Criterion

With the assistance of the traffic estimate dataset, hourly average traffic information of all of the major and minor road segments of New York City is determined. This dataset includes the traffic estimates from 2010 to 2013, of the city based on the estimation of approximately 700 million taxi trips [40,41]. For the computation of *traffic* criteria, this dataset is exploited.

While observing all of the criteria, traffic is the second most important criterion that affects the gas station sales. In Figure 3, the locations are highlighted, where per hour traffic estimates are performed in New York City. The estimated traffic within radius r_t of each $l \in L$ is computed to determine the traffic near the candidate locations. Here, the assumption is that T represents the set of points where traffic estimation for NYC is available and t represents the point at which the traffic is estimated. The traffic estimation within radius r_t of each $l \in L$ is performed using Equation (2).

$$|\{t \in T : dist(t,l) < r_t\}| \quad (2)$$

Figure 3. NYC traffic estimate.

5.1.3. FourSquare Check-Ins Dataset and the Area-Popularity Criterion

The area-popularity criterion can be easily found with a location-aware social networking application. Here, FourSquare is exploited to gather user check-in details at various business attractions to find out their popularity. The FourSquare dataset for this research work consists of check-in information within New York City from 12 April 2012 to 16 February 2013 (approx. 10 months). A total of 227,428 check-ins in New York City are contained in this data set and are shown in Figure 4. There is a time stamp, GPS coordinates, and some semantic meaning (expressed with the fine-grained venue-categories) connected to every check-in [42]. The area *popularity* factor is computed with the help of this check-in dataset.

Another significant factor for picking out a suitable location to open a new gas station is area popularity. The potential number of customers can be estimated with this criteria. A distance function, similar to that given in Equation (1), is used for finding the area popularity. Here, C represents the set of all check-ins in the dataset and $c \in C$ denotes an individual check-in instance at some geographical location. Hence, with the help of Equation (3), the total number of check-ins of each candidate location $l \in L$ within the radius r_c are estimated.

$$|\{c \in C : dist(c, l) < r_c\}| \tag{3}$$

5.1.4. Parking Lot Datasets and the Vehicle-Owner Criterion

The NYC Open Data [39] also provides New York City parking lot data. A total of 20,715 parking lots in New York City are contained in this dataset along with their geographical location and the size (in terms of area). In Figure 5, polygons are used to show the location and the area of the parking lots. The criterion vehicle owners, i.e., the number of people who own a vehicle in the surroundings of the candidate gas stations are estimated by employing this dataset.

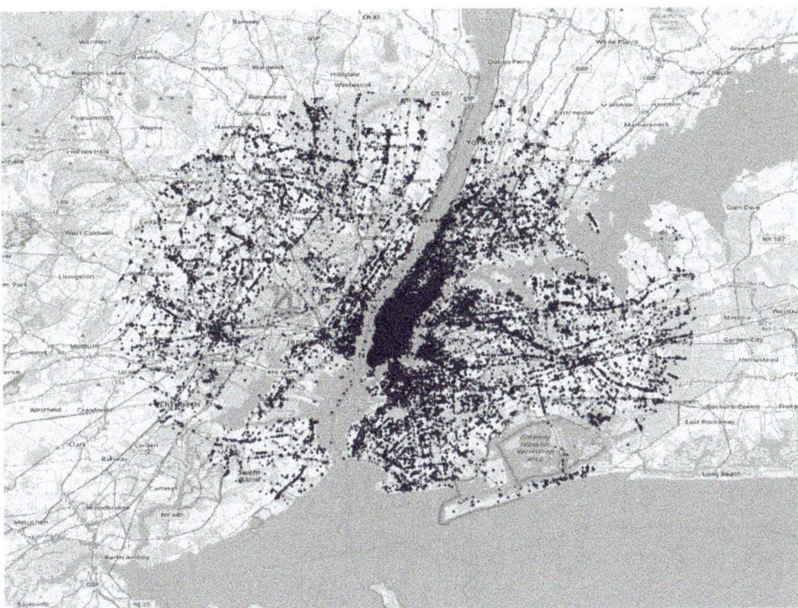

Figure 4. NYC FourSquare check-ins.

Figure 5. NYC parking lots.

It is obvious that a gas station is accessed by vehicle owners only. Therefore, this criterion plays a significant role in this research. Although it is challenging to determine the actual number of people who own vehicles in New York City, the NYC Open Data provides parking lots details that can be processed for estimating this factor. An estimated number of parking slots per area can be obtained by dividing the area according to the parking standards [43,44]. Suppose that V represents the set of all parking lots; then, the function given in Equation (4) is employed to determine the approximate number of owned vehicles that are in close proximity to a candidate location $l \in L$ with the help of the approximate number of parking slots that are located within distance r_v of l.

$$|\{v \in V : dist(v, l) < r_v\}| \quad (4)$$

5.2. Criteria Computation

In this step, the competitors, traffic, popularity, and vehicle owner criteria are considered to select an optimal gas station site. Once the set of candidate gas station sites is provided, the very first step is to compute the criteria. Second, the candidate sites are evaluated with the AHP/TOPSIS technique to identify the ideal site for a new gas station. In Figure 6, red stars represent the candidate gas station sites (provided by user) and green circles explicitly provide the location of existing gas station sites in New York City.

For the calculation of criteria traffic, popularity, vehicle owners, and competitors, the distance functions given in Section 5.1 and the radius values r_t: 1000, r_c: 1000, r_v: 5000, and r_g: 3000 m are used, respectively. Table 2 shows the candidate sites' computed criteria.

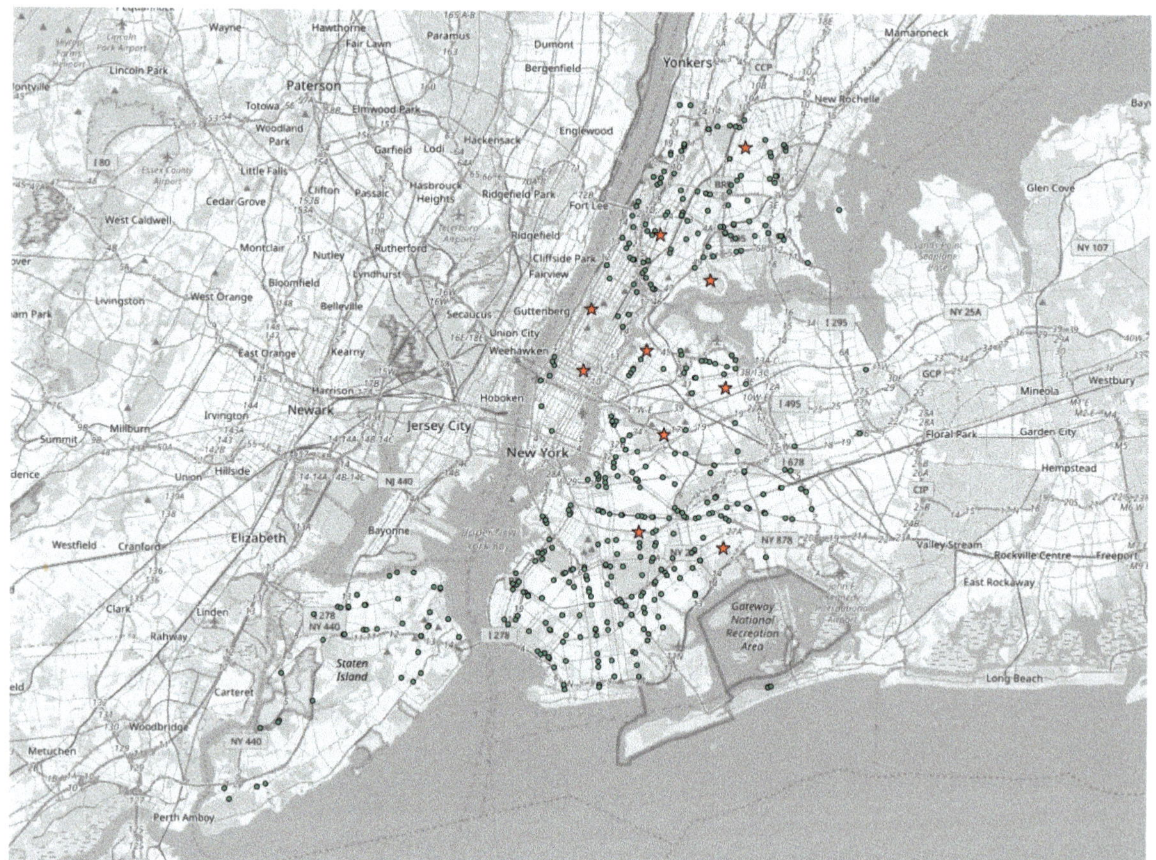

Figure 6. NYC gas stations and candidate locations.

Table 2. Candidate sites factor computation.

CID	Alternatives	Factors			
		Traffic	Popularity	Vehicle Owners	Competitors
		1000 m	1000 m	5000 m	3000 m
1	40.8839, −73.8561	2.91667	178	28,029	23
2	40.8095, −73.8807	3.89881	79	56,643	7
3	40.8347, −73.9177	228.345	631	42,954	32
4	40.7931, −73.9675	16,170.9	1171	19,815	7
5	40.7586, −73.9731	73,897.9	15,316	25,402	7
6	40.7699, −73.9271	2624.86	1276	47,071	13
7	40.7496, −73.8692	416.202	200	51,613	16
8	40.7231, −73.914	507.357	33	44,028	10
9	40.6686, −73.9321	110.815	340	30,995	38
10	40.6599, −73.8701	2.57738	122	33,898	20

5.3. Criteria Weight Computation via AHP

In computing the factors' weights, the very first step is the hierarchical model construction. The AHP hierarchical model consists of the goal at the top or at the root level. The criteria are placed at the intermediate level, whereas the candidates or alternatives are placed at the bottom. The AHP hierarchical model for our optimal site selection problem is shown in Figure 7.

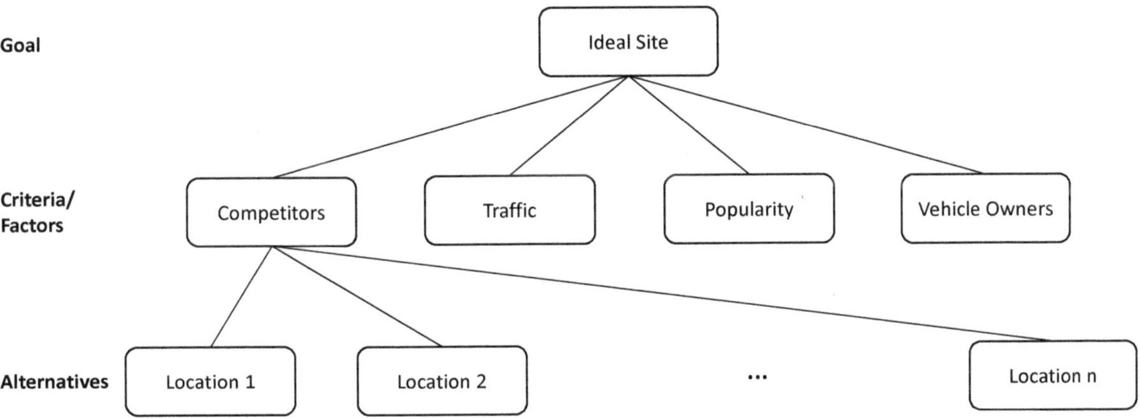

Figure 7. AHP hierarchical model.

The next step in the AHP is the construction of a pairwise comparison matrix **A** to compute the criteria priorities/weights. The matrix **A** is a $m \times m$ real matrix, where m is the number of evaluation criteria/factors considered. Each entry a_{jk} of the matrix **A** represents the importance of the jth criterion relative to the kth criterion. If $a_{jk} > 1$, then the jth criterion is more important than the kth criterion, while if $a_{jk} < 1$, then the jth criterion is less important than the kth criterion. If two criteria have the same importance, then the entry a_{jk} is 1. The entries a_{jk} and a_{kj} satisfy the following constraint [11]:

$$a_{jk}.a_{kj} = 1$$

The following assumptions for the four factors are derived based on the pairwise comparison matrix. Table 3 shows the pairwise comparison matrix.

- The most significant factor is the number of competitors nearby.

- The second most significant factor is nearby traffic, followed by popularity of the area.
- The least important factor found in this study is vehicle ownership in the vicinity.

Table 3. Pairwise comparison matrix.

Factors	Competitors	Traffic	Popularity	VOwners
Competitors	1	5	6	8
Traffic	0.2	1	5	7
Popularity	0.1667	0.2	1	3
Vehicle Owners	0.125	0.1429	0.3333	1

The factors' weight is computed after normalization of the pairwise comparison matrix. The normalized pairwise comparison matrix $\boldsymbol{A_{norm}}$ is derived by making the sum of the entries in each column equal to 1, i.e., each entry $\overline{a}_{jk}$ of the matrix $\boldsymbol{A_{norm}}$ is computed as

$$\overline{a}_{jk} = \frac{a_{jk}}{\sum_{l=1}^{m} a_{lk}}$$

Ultimately, the criteria/factors weight vector $\mathbf{w}$ is obtained by averaging the entries on each row of the matrix $\boldsymbol{A_{norm}}$ and can be computed as follows:

$$w_j = \frac{\sum_{l=1}^{m} \overline{a}_{jl}}{m}$$

Table 4 shows the normalized pairwise comparison matrix with the weights of the criteria. It is necessary to calculate the consistency of the pairwise comparison matrix; therefore, it is performed after criteria weight computation. The acceptance of the criteria weight $\mathbf{w}$ depends upon the consistency ratio, which must be less than 0.1; otherwise, it is assumed that the selection of comparison matrix values are not consistent. In this situtaion, the values of the pairwise comparison matrix need to be reallocated. In order for the matrix to be consistent, please refer to [9,25].

Table 4. Normalized pairwise comparison matrix with factors weights.

Factors	Competitors	Traffic	Popularity	V Owners	Factors Weight
Competitors	0.676691	0.779221	0.529411	0.473684	**0.614752121**
Traffic	0.135338	0.155844	0.352941	0.315789	**0.239978288**
Popularity	0.112781	0.038961	0.088235	0.157895	**0.099468256**
V Owners *	0.07518	0.025974	0.029411	0.052631	**0.045801335**

* V Owners: Vehicle owners.

5.4. Ranking the Alternatives Using TOPSIS

Once the criteria weights are determined with the help of AHP [11], TOPSIS is utilized to assign ranks to each alternative (gas station candidate sites). According to the researchers in [10], TOPSIS is a multi-criteria decision analysis method (MCDM), which assists in selecting the best option among a finite set of decision alternatives. A step-by-step procedure is shown below for computing the ranking of the alternatives.

Step 1: This step makes use of TOPSIS to make an $n \times m$ evaluation matrix $\boldsymbol{E}$ consisting of n alternatives and m criteria. Without losing generality, 10 alternatives (gas station candidate sites as shown in Figure 6) and 4 criteria are chosen. Table 5 presents a TOPSIS evaluation matrix of the 10 candidate sites and 4 factors.

Step 2: In this step, the matrix $\boldsymbol{E}$ is normalized to form the matrix $\boldsymbol{E_{norm}}$, where each entry $\bar{e}_{ij}$ of $\boldsymbol{E_{norm}}$ is computed as

$$\bar{e}_{ij} = \frac{e_{ij}}{\sqrt{\sum_{k=1}^{n} e_{kj}^2}}$$

where $i = 1, 2, ..., n$ and $j = 1, 2, ..., m$.

Step 3: In this step, the weighted normalized decision matrix $\boldsymbol{E_{weighted}}$ is obtained by multiplying the criteria weights w_j (computed in Section 5.3) to the corresponding criteria values. Hence, each entry e_{ij}^{w} of $\boldsymbol{E_{weighted}}$ is computed as $\bar{e}_{ij}.w_j$. The weighted normalized decision matrix is represented by the grey columns in Table 5.

Table 5. Weighted normalized decision matrix.

ID	Candidates	Traffic	Popularity	Vehicle Owners	Competitors
1	40.8839, −73.8561	9.24677×10^{-6}	1.14721×10^{-3}	1.02115×10^{-2}	2.21658×10^{-1}
2	40.8095, −73.8807	1.23605×10^{-5}	5.09158×10^{-4}	2.06362×10^{-2}	6.74612×10^{-2}
3	40.8347, −73.9177	7.23926×10^{-4}	4.06682×10^{-3}	1.56490×10^{-2}	3.08394×10^{-1}
4	40.7931, −73.9675	5.12669×10^{-2}	7.54714×10^{-3}	7.21901×10^{-3}	6.74612×10^{-2}
5	40.7586, −73.9731	2.34279×10^{-1}	9.87122×10^{-2}	9.25446×10^{-3}	6.74612×10^{-2}
6	40.7699, −73.9271	8.32164×10^{-3}	8.22387×10^{-3}	1.71489×10^{-2}	1.25285×10^{-1}
7	40.7496, −73.8692	1.31949×10^{-3}	1.28901×10^{-3}	1.88036×10^{-2}	1.54197×10^{-1}
8	40.7231, −73.914	1.60848×10^{-3}	2.12686×10^{-4}	1.60403×10^{-2}	9.63731×10^{-2}
9	40.6686, −73.9321	3.51319×10^{-4}	2.19131×10^{-3}	1.12921×10^{-2}	3.66218×10^{-1}
10	40.6599, −73.8701	8.17112×10^{-6}	7.86295×10^{-4}	1.23497×10^{-2}	1.92746×10^{-1}
	V_j^+	2.34279×10^{-1}	9.87122×10^{-2}	2.06362×10^{-2}	6.74612×10^{-2}
	V_j^-	8.17112×10^{-6}	2.12686×10^{-4}	7.21901×10^{-3}	3.66218×10^{-1}

Step 4: Next, we determine the worst alternative (V_j^-) and the best alternative (V_j^+) for each column in $\boldsymbol{E_{weighted}}$ as follows:

$$V_j^- = \{(max_i\, e_{ij}^{w} | j \in J_-), (min_i\, e_{ij}^{w} | j \in J_+)\}$$

$$V_j^+ = \{(min_i\, e_{ij}^{w} | j \in J_-), (max_i\, e_{ij}^{w} | j \in J_+)\}$$

where $i = 1, 2, ..., n$, $J_+ = \{j = 1, 2, ..., m\}$ is associated with the criteria having a positive impact and $J_- = \{j = 1, 2, ..., m\}$ is associated with the criteria having a negative impact.

Step 5: Next, we need to compute the Euclidean distance ($L^2 -$ distance) between the target alternative i and the best alternative V_j^+ and between the target alternative i and the worst alternative V_j^-, denoted by S_i^+ and S_i^-, respectively, and given as follows:

$$S_i^+ = \sqrt{\sum_{j=1}^{m} (e_{ij}^{w} - V_j^+)^2}, i = 1, 2, ..., n$$

$$S_i^- = \sqrt{\sum_{j=1}^{m} (e_{ij}^{w} - V_j^-)^2}, i = 1, 2, ..., n$$

Step 6: Finally, the performance score (P_i) of each ith alternative is computed using the following equation.

$$P_i = \frac{S_i^-}{S_i^- + S_i^+}$$

By comparing the P_i values, the ranking of alternatives is determined: the higher the value, the better the rank. Table 6 shows the S_i^-, S_i^+, and P_i values along with the candidates/alternatives ranking.

From the Table 6, it is deduced that candidate having ID 5 is ranked first; therefore, a new gas station can be opened at this optimal site, followed by candidate IDs (CID) 4 and 2. The performance scores suggest that CID 9 must be avoided because it ranks worst among all of the candidate sites.

Table 6. Candidate S_i^-, S_i^+, and P_i values and their ranking.

ID	Candidates	S_i^+	S_i^-	P_i	Rank
1	40.8839, −73.8561	0.297131401	0.144593761	0.327338747	8
2	40.8095, −73.8807	0.254017926	0.299058138	0.540717918	3
3	40.8347, −73.9177	0.348682791	0.05856651	0.143809971	9
4	40.7931, −73.9675	0.204902022	0.303211003	0.596739284	2
5	40.7586, −73.9731	0.011381735	0.392231042	0.97180036	1
6	40.7699, −73.9271	0.250201973	0.241413717	0.491061864	5
7	40.7496, −73.8692	0.266998664	0.212344026	0.442990016	6
8	40.7231, −73.914	0.254352423	0.269993794	0.514915118	4
9	40.6686, −73.9321	0.391639917	0.004541242	0.011462539	10
10	40.6599, −73.8701	0.283262679	0.173548531	0.37991303	7

6. Experiments

In this section, we evaluate the effectiveness of the proposed AHP/TOPSIS approach to identify an ideal business location. As mentioned earlier, ideal business location in this work means the location that can attract the maximum number of customers. For the evaluation in this section, sensitivity measurement is used, which can be defined as the proportion of positives (correct results) that are correctly identified, also known as True Positive Rate (TPR). Let the TP and FN denote the number of true positives and false negatives, respectively; then, the TPR is given by the following:

$$TPR = \frac{TP}{TP + FN} \tag{5}$$

To evaluate the proposed method effectiveness, an NYC convenience store dataset is used to identify the popularity and/or success of a convenience store, and its visitor count is used.

6.1. Datasets and Experimental Setup

For the evaluation, we made use of NYC conveninece store checkin data available from FourSquare [45]. To obtain convenience stores' location data and their visitor counts from FourSquare, the developers' places API was used. API calls to APIs can be broken down into two categories: regular and premium. Regular API calls only return basic information including the venue location, category, and a venue ID. Premium API calls return rich content including the number of visitors. In order to obtain convenience store data, premium API calls were used as we were interested in convenience store's visitor count in addition to its location information.

For the sake of evaluation, we ranked the NYC convenience stores in descending order with respect to their visitor count and used it as *ground truth*. Although the visitor count is reliable, the duration of the visitor count is not known. We then used our proposed AHP/TOPSIS approach to rank the NYC convenience stores. The obtained ranking is compared against the ground truth, and the TPR is computed.

For the convenience stores' ideal location, we identified the following five criteria based on the intuition that most of the convenience stores' customers come from these locations:

1. TStations: Transportation stations (train, metro, bus, stations, etc.);

2. Buildings: Buildings including residential, commercial, etc.;
3. EVenues: Entertainment venues (museums, movie theaters, stadiums, etc.);
4. Shops: Other shops in the vicinity; and
5. PPlaces: Professional places (convention centers, medical centers, factories, etc.)

The data (venue information) related to the criteria *TStations*, *EVenues*, *Shops*, and *PPlaces* were obtained from the FourSquare developers' places API [46]. Using the API, up to date check-in information was obtained for over 62 million global venues (As of 31 August 2020). The data related to criteria *Buildings* was obtained from the NYC buildings footprint data [47]. Building footprints represent the full perimeter outline of each building as viewed from directly above. Besides the perimeter, other useful attributes of this dataset included ground elevation at building base, roof height above ground elevation, construction year, and feature type. The Buildings dataset consists of more than 1 million NYC building information including residential, commercial, and government buildings.

6.2. Evaluation

As discussed in Section 5.2, we computed the criteria values for the five criteria mentioned in Section 6.1. The radius values 1000, 300, 1000, 1000, and 1000 m were used for the criteria TStations, Buildings, EVenues, Shops, and PPlaces, respectively. The reason for using a smaller radius for the criterion Buildings compared to the other criteria is that the Buildings dataset is quite large and we obtain a uniform and a large number of buildings for each candidate for radius 1000 m. Thus, limiting the radius to a smaller value, in this case to 300 m, help us identify the nearby population of a convenience store. The same is not true for the other criteria, i.e., TStations, EVenues, Shops, and PPlaces. Thus radius values of 1000 m are used for them.

Table 7 shows the pairwise comparison matrix for the five criteria. In the criteria computation, we assume that the criterion TStations, i.e., transportation stations, is the most important criteria as a large number of people visit convenience stores during their commute or travel. The second most important criterion that we identified is buildings. A large number of buildings around a convenience store means a large number of people either living or working there. Criteria TStations and Buildings are followed by criteria EVenues, Shops, and PPlaces, which are comparatively less significant compared to the first two criteria. In addition to the pairwise comparison matrix, Table 7 shows the computed criteria weights. The details of its computation are discussed in Section 5.2.

Table 7. Pairwise comparison matrix and computed criteria weights.

Criteria	TStations	Buildings	EVenues	Shops	PPlaces	Weight
TStations	1	3	7	9	7	0.495486996
Buildings	0.3333	1	7	9	7	0.327448046
EVenues	0.1428	0.1428	1	3	1	0.071501752
Shops	0.1111	0.1111	0.3333	1	0.3333	0.034061454
PPlaces	0.1428	0.1428	1	3	1	0.071501752

Tables 8 and 9 show the top 20 stores ranked based on FourSquare visitor count (ground truth) and the top 20 stores ranked by our AHP/TOPSIS approach, respectively. Figure 8 shows the NYC map containing the convenience stores in Tables 8 and 9. Bold tuples in Table 9 are the ones ranked by the ground truth in Table 8 as well. Based on Tables 8 and 9, we computed the True Positive Rate (TPR). We counted the records predicted by AHP/TOPSIS, i.e., the record that is present in Table 9 is a True Positive (TP) if it also appears in the ground truth table i.e., Table 8, irrespective of its rank. On the other hand, a record is counted as a False Negative (FN) if it is predicted by AHP/TOPSIS but it does not appear in Table 8. Hence, the TPR for the criteria weights computed in Table 7 is given as follows:

$$TPR = \frac{TP}{TP + FN} = \frac{11}{20} = 0.55$$

The TPR is heavily dependent on the criteria weights, and the derivation of the right criteria weights is important to obtain an optimal or desired prediction. To prove this, we performed experiments with a number of manual weight assignments to the criteria vector [TStations, Buildings, EVenues, Shops, PPlaces] as follows:

- uniform = [0.2, 0.2, 0.2, 0.2, 0.2]
- increasing = [0.05, 0.1, 0.2, 0.3, 0.35]
- decreasing = [0.35, 0.3, 0.2, 0.1, 0.05]
- oneCriteriaZero = [0.4, 0.3, 0.2, 0.1, 0]

For the random weight vectors uniform, increasing, decreasing, and oneCriteriaZero above, we obtained the TPRs 0.25, 0.25, 0.4, and 0.35, respectively. The TPRs are far lower than the one obtained using carefully derived weights, i.e., 0.55. Figure 9 shows the placement of ground truth convenience stores (green triangles) and the AHP/TOPSIS-predicted convenience stores using the manual weights (red circles). As can be observed from the obtained TPR values of the different weight vectors, the AHP/TOPSIS approach is heavily dependent on the criteria weight computation.

Table 8. Ground truth—stores ranked based on FourSquare visitor count.

Rank	Store ID	Store Name
1	5cd15dc59d7468003903fe8a	Amazon Go
2	5266a90711d23056d7c6dc67	Bread & Butter
3	4e034363b61ce80e5d67d09f	Duane Reade
4	59ceacc02955135d6151aef3	CVS pharmacy
5	4b0b3438f964a520982e23e3	Walgreens
6	4cdc1f63df986ea8550fce16	CVS pharmacy
7	4ca4e5b214c33704a852b13b	Duane Reade
8	5678271c498ed123ac7fc1e0	CVS Pharmacy
9	4c545b1b479fc9280e360293	Duane Reade
10	58768de8f2299508aa731849	CVS pharmacy
11	4b61db21f964a52060272ae3	Duane Reade
12	4b130a80f964a520239323e3	Duane Reade
13	4b95bf56f964a520cdb134e3	Duane Reade
14	5b92aaf3033693002c2af840	Open Market
15	4b574341f964a5200e2e28e3	Duane Reade
16	4aca8742f964a52041c220e3	Duane Reade
17	4ae1ed7cf964a520d58821e3	Duane Reade
18	4fe4f632a17c9aa22d55209f	Duane Reade
19	4a7f68e6f964a520eff31fe3	Duane Reade
20	4a9e7fadf964a520503a20e3	Duane Reade

Table 9. AHP/TOPSIS—stores ranked obtained using the AHP/TOPSIS approach.

Rank	Store ID	Store Name	Ground Truth Ranking *
1	**4b0b3438f964a520982e23e3**	**Walgreens**	**5**
2	**59ceacc02955135d6151aef3**	**CVS pharmacy**	**4**
3	**4b61db21f964a52060272ae3**	**Duane Reade**	**11**
4	4ad3f4e5f964a52011e720e3	Duane Reade	NR
5	**4e034363b61ce80e5d67d09f**	**Duane Reade**	**3**
6	4a9c8a06f964a520813720e3	Duane Reade	NR
7	**5b92aaf3033693002c2af840**	**Open Market**	**14**
8	**4fe4f632a17c9aa22d55209f**	**Duane Reade**	**18**
9	59a7303f8d0a5343731f326c	7 Eleven	NR
10	**4aca8742f964a52041c220e3**	**Duane Reade**	**16**
11	4d6b8ebe56cc6a3175a95eac	7-Eleven	NR
12	4adb7ccdf964a520152821e3	Duane Reade	NR
13	5d4117543c3cf7000794b01b	CVS Pharmacy	NR
14	4e7a39122271920e476b47e3	7-Eleven	NR
15	**5266a90711d23056d7c6dc67**	**Bread & Butter**	**2**
16	4adf4335f964a520df7821e3	Duane Reade	NR
17	**5cd15dc59d7468003903fe8a**	**Amazon Go**	**1**
18	**4ae1ed7cf964a520d58821e3**	**Duane Reade**	**17**
19	**4a7f68e6f964a520eff31fe3**	**Duane Reade**	**19**
20	4ba57508f964a520160939e3	Duane Reade	NR

* NR: Not Ranked.

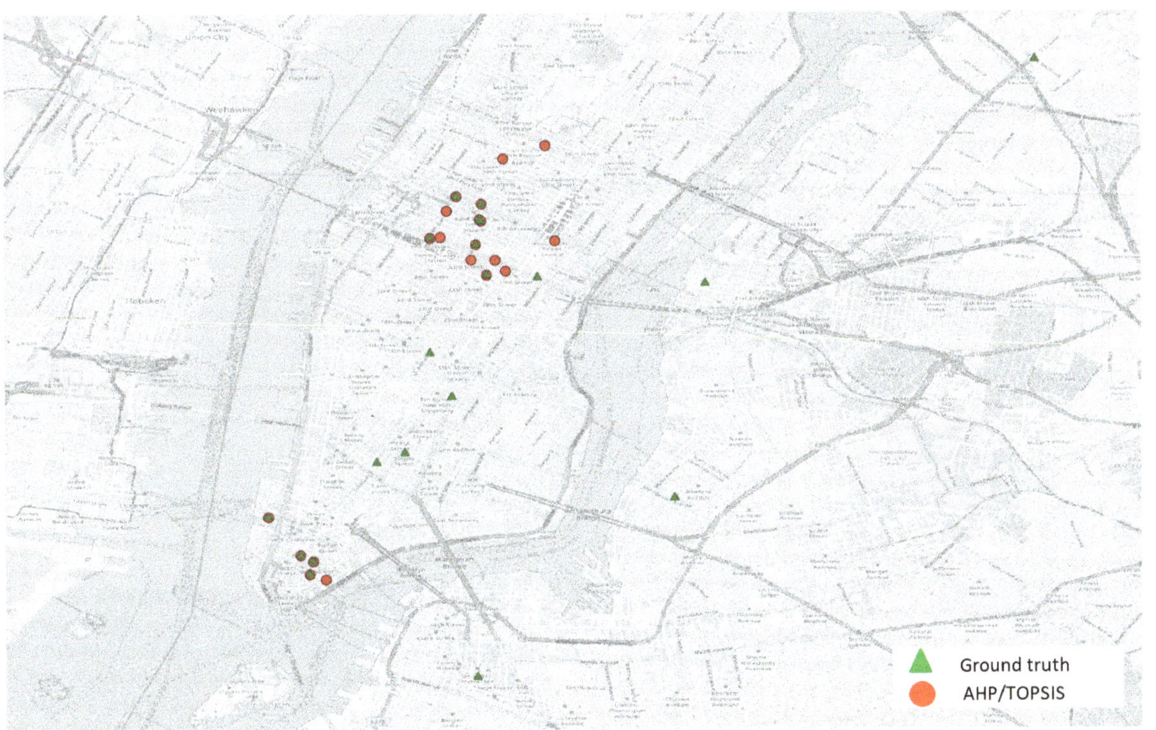

Figure 8. Convenience store placement: ground truth vs. AHP/TOPSIS.

(a) Uniform

(b) Increasing

(c) Decreasing

(d) One Criteria Zero

Figure 9. Manual weight assignments (ground truth convenience stores (green triangles) and AHP/TOPSIS convenience stores (red circles)).

7. Discussion

By looking at the result, it seems that the AHP/TOPSIS method is not as effective as we could obtain only 55% correct results (TP) compared to the ground truth. However, we would like to argue that ground truth raking is based on the number of visitors, which can be biased. For instance, the number of visitors depends on several factors besides the five criteria, i.e., TStations, Buildings, EVenues, Shops, and PPlaces, which we considered in our AHP/TOPSIS approach. Thus, in order to improve the accuracy of the AHP/TOPSIS model, more data sets are needed. For instance, stores' daily sales, pricing policy, product line, timed sales, special sales, etc. are very important criteria that play important roles in attracting visitors and in improving daily sales. In fact, daily sales is a better criteria to rank stores than the number of visitors used in this study. However, the data related to such criteria is very difficult to obtain if not impossible because of stores' privacy policies. We strongly believe that the accuracy of the results can be significantly improved with the combination of the right criteria and respective datasets.

By analyzing the top 20 stores in Figure 8, one can observe that the proposed AHP/TOPSIS approach identified convenience stores mainly at the center of the NYC Manhattan area, which makes sense as it is the most crowded area with a lot of residential and commercial buildings, transportation stations, shops/markets, and entertainment venues. Since in the experiments, the highest weights were allocated to the TStations and Buildings criteria, AHP/TOPSIS identified business locations at a crowded part of NYC. The most important step in the AHP/TOPSIS approach is the identification of the set of important criteria and the derivation of their weights with the help of a pairwise comparison matrix. The accuracy of the the AHP/TOPSIS approach is heavily dependent on the criteria weight computation.

Section 6.2 shows that, for the random weights assignment, i.e., the weight vectors uniform, increasing, decreasing, and oneCriteriaZero, we obtained the TPRs 0.25, 0.25, 0.4, and 0.35, respectively, which are far lower than the one obtained using the careful derivation of criteria weights. This proves that the AHP/TOPSIS approach is sensitive to criteria weight assignments.

8. Conclusions and Future Work

In this paper, the challenge of selecting an optimal site to open a commercial place is addressed. It is evident with the support of related research that an ideal site selection to open a new business requires thorough study of various factors and criteria. This study puts forward an AHP/TOPSIS-based hybrid solution for this problem, where AHP and TOPSIS are two state-of-the-art multi-criteria decision-making approaches. The proposed approach helps to identify the best alternative among the given candidates for a commercial opening while minimizing the computational complexity and reducing the manual effort required. The classic MCDM approach AHP becomes computationally expensive, i.e., requires a large number of comparison matrices to be constructed in the presence of a large number of alternatives/candidates. Thus, the proposed AHP/TOPSIS approach is particularly useful to solve this issue and can come up with the best alternative using only two matrices: one for riteria weight computation and the other for alternatives ranking. The applicability of the proposed approach is demonstrated with the help of a detailed step-by-step case study and to evaluate the effectiveness of the proposed approach, and an experimental evaluation to identify an optimal location for a convenience store is presented. In the presented case study, a nexus of four criteria was considered, including competitors, traffic, popularity, and vehicle owners, where the criteria values were computed with the assistance of real GeoSpatial data. In contrast, for the evaluation, we made use of five criteria, namely, TStations, Buildings, EVenues, Shops, and PPlaces, which were based on real data. This study claims that the proposed mechanism is flexible in a sense that any number of criteria can be exploited for ranking the candidate sites. The results of the evaluation suggests that the proposed approach is highly dependent on the criteria weights and that the derivation of the right weight matrix is important to obtain a correct prediction. Since the derivation of criteria weight is a heuristic approach, it is both flexible and error-prone, flexible in the sense that it enables users to give preference to one of more criteria of their choice and error-prone because of manual computation of a weight matrix. In the future, this work will be extended to identify an optimal route for mobile businesses.

Author Contributions: Conceptualization, S.A.S., M.M. and K.-S.K.; methodology, S.A.S. and M.M.; software, S.A.S.; validation, S.A.S.; formal analysis, S.A.S.; investigation, S.A.S. and M.M.; resources, S.A.S.; data curation, S.A.S.; writing—original draft preparation, S.A.S. and M.M.; writing—review and editing, S.A.S. and M.M.; visualization, S.A.S.; supervision, S.A.S.; project administration, S.A.S.; funding acquisition, K.-S.K. All authors have read and agreed to the published version of the manuscript.

Funding: This research was funded by New Energy and Industrial Technology Development Organization (NEDO) grant number JPNP18010.

Institutional Review Board Statement: Not applicable.

Informed Consent Statement: Not applicable.

Acknowledgments: This article is based on results obtained from a project, JPNP18010, commissioned by the New Energy and Industrial Technology Development Organization (NEDO).

Conflicts of Interest: The authors declare no conflict of interest.

References

1. Dodgson, J.; Spackman, M.; Pearman, A.; Phillips, L. *Multi-Criteria Analysis: A Manual*; Department for Communities and Local Government: London, UK, 2009.
2. Triantaphyllou, E. *Multi-Criteria Decision Making Methods: A Comparative Study*; Springer: Berlin/Heidelberg, Germany, 2000; Volume 44. [CrossRef]
3. Onut, S.; Efendigil, T.; Kara, S.S. A combined fuzzy MCDM approach for selecting shopping center site: An example from Istanbul, Turkey. *Expert Syst. Appl.* **2010**, *37*, 1973–1980. [CrossRef]
4. GarcÃa, J.; Alvarado, A.; Blanco, J.; Jiménez, E.; Maldonado, A.; Cortés, G. Multi-attribute evaluation and selection of sites for agricultural product warehouses based on an Analytic Hierarchy Process. *Comput. Electron. Agric.* **2014**, *100*, 60–69. [CrossRef]
5. Aliniai, K.; Yarahmadi, A.; Zarin, J.; Yarahmadi, H.; Lak, S. Parking Lot Site Selection: An Opening Gate Towards Sustainable GIS-based Urban Traffic Management. *J. Indian Soc. Remote. Sens.* **2015**, *43*. [CrossRef]
6. Karamshuk, D.; Noulas, A.; Scellato, S.; Nicosia, V.; Mascolo, C. Geo-spotting: Mining Online Location-based Services for Optimal Retail Store Placement. In Proceedings of the 19th ACM SIGKDD International Conference on Knowledge Discovery and Data Mining, KDD '13, Chicago, IL, USA, 11–14 August 2013; ACM: New York, NY, USA, 2013; pp. 793–801. [CrossRef]
7. Semih, T.; Seyhan, S. A Multi-Criteria Factor Evaluation Model For Gas Station Site Selection. *J. Glob. Manag.* **2011**, *2*, 12–21.
8. Athey, S.; Blei, D.; Donnelly, R.; Ruiz, F.; Schmidt, T. Estimating Heterogeneous Consumer Preferences for Restaurants and Travel Time Using Mobile Location Data. *arXiv* **2018**, arXiv:1801.07826.
9. Saaty, R. The analytic hierarchy process—What it is and how it is used. *Math. Model.* **1987**, *9*, 161–176. [CrossRef]
10. Hwang, C.L.; Yoon, K. *Multiple Attribute Decision Making, Methods and Applications A State-of-the-Art Survey*; Lecture Notes in Economics and Mathematical Systems; Springer: Berlin/Heidelberg, Germany, 1981.
11. Saaty, T.L. Decision making with the analytic hierarchy process. *Int. J. Serv. Sci.* **2008**, *1*, 83–98. [CrossRef]
12. Sangiorgio, V.; Uva, G.; Fatiguso, F. Optimized AHP to Overcome Limits in Weight Calculation: Building Performance Application. *J. Constr. Eng. Manag.* **2018**, *144*, 04017101. [CrossRef]
13. Sangiorgio, V.; Martiradonna, S.; Fatiguso, F.; Lombillo, I. Augmented reality based-decision making (AR-DM) to support multi-criteria analysis in constructions. *Autom. Constr.* **2021**, *124*, 103567. [CrossRef]
14. Li, X.; Wang, K.; Liu, L.; Xin, J.; Yang, H.; Gao, C. Application of the Entropy Weight and TOPSIS Method in Safety Evaluation of Coal Mines. *Procedia Eng.* **2011**, *26*, 2085–2091. [CrossRef]
15. Shaikh, S.A.; Memon, M.A.; Prokop, M.; Kim, K. An AHP/TOPSIS-Based Approach for an Optimal Site Selection of a Commercial Opening Utilizing GeoSpatial Data. In Proceedings of the 2020 IEEE International Conference on Big Data and Smart Computing (BigComp), Busan, Korea, 19–22 February 2020; pp. 295–302.
16. Bean, J. Analyzing and Predicting Starbucks' Location Strategy. Available online: https://towardsdatascience.com/analyzing-and-predicting-starbucks-location-strategy-3c5026d31c21 (accessed on 12 June 2019).
17. Turhan, G.; Akalın, M.; Zehir, C. Literature Review on Selection Criteria of Store Location Based on Performance Measures. *Procedia-Soc. Behav. Sci.* **2013**, *99*, 391–402. [CrossRef]
18. Vahidnia, M.H.; Alesheikh, A.A.; Alimohammadi, A. Hospital site selection using fuzzy AHP and its derivatives. *J. Environ. Manag.* **2009**, *90*, 3048–3056. [CrossRef] [PubMed]
19. Wang, G.; Qin, L.; Li, G.; Chen, L. Landfill site selection using spatial information technologies and AHP: A case study in Beijing, China. *J. Environ. Manag.* **2009**, *90*, 2414–2421. [CrossRef] [PubMed]
20. Awasthi, A.; Chauhan, S.; Goyal, S. A multi-criteria decision making approach for location planning for urban distribution centers under uncertainty. *Math. Comput. Model.* **2011**, *53*, 98–109. [CrossRef]
21. Prakash, C.; Barua, M. Integration of AHP-TOPSIS method for prioritizing the solutions of reverse logistics adoption to overcome its barriers under fuzzy environment. *J. Manuf. Syst.* **2015**, *37*, 599–615. [CrossRef]
22. Tyagi, M.; Kumar, P.; Kumar, D. A Hybrid Approach using AHP-TOPSIS for Analyzing e-SCM Performance. In Proceedings of the 12th Global Congress on Manufacturing and Management GCMM-2014, Vellore, India, 8–10 December 2014. [CrossRef]
23. Supraja, S.; Kousalya, P. A comparative study by AHP and TOPSIS for the selection of all round excellence award. In Proceedings of the 2016 International Conference on Electrical, Electronics, and Optimization Techniques (ICEEOT), Chennai, India, 3–5 March 2016, pp. 314–319.
24. Zafri, N.; Sameen, I.; Jahangir, A.; Tabassum, N.; Hasan, M.M.U. A multi-criteria decision-making approach for quantification of accessibility to market facilities in rural areas: An application in Bangladesh. *GeoJournal* **2020**, 1–17. [CrossRef]
25. Önder, E.; Dag, S. Combining Analytical Hierarchy Process and Topsis Approaches for Supplier Selection in a Cable Company. *J. Bus. Econ. Financ.* **2013**, *2*, 56–74.
26. Jozaghi, A.; Alizadeh, B.; Hatami, M.; Flood, I.; Khorrami, M.; Khodaei, N.; Ghasemi Tousi, E. A Comparative Study of the AHP and TOPSIS Techniques for Dam Site Selection Using GIS: A Case Study of Sistan and Baluchestan Province, Iran. *Geosciences* **2018**, *8*, 494. [CrossRef]
27. Sangiorgio, V.; Uva, G.; Aiello, M.A. A multi-criteria-based procedure for the robust definition of algorithms aimed at fast seismic risk assessment of existing RC buildings. *Structures* **2020**, *24*, 766–782. [CrossRef]
28. Abdel-Basset, M.; Gamal, A.; Chakrabortty, R.K.; Ryan, M.J. Evaluation of sustainable hydrogen production options using an advanced hybrid MCDM approach: A case study. *Int. J. Hydrogen Energy* **2021**, *46*, 4567–4591. [CrossRef]

29. Sedghiyan, D.; Ashouri, A.; Maftouni, N.; Xiong, Q.; Rezaee, E.; Sadeghi, S. Prioritization of renewable energy resources in five climate zones in Iran using AHP, hybrid AHP-TOPSIS and AHP-SAW methods. *Sustain. Energy Technol. Assess.* **2021**, *44*, 101045. [CrossRef]
30. Cinar, U.; Cebi, S. A hybrid risk assessment method for mining sector based on QFD, fuzzy logic and AHP. In Proceedings of the International Conference on Intelligent and Fuzzy Systems, Istanbul, Turkey, 23–25 July 2019; Springer: Cham, Switzerland, 2019, pp. 1198–1207.
31. Keskin, B.; Köksal, C.D. A hybrid AHP/DEA-AR model for measuring and comparing the efficiency of airports. *Int. J. Product. Perform. Manag.* **2019**, *68*, 524–541. [CrossRef]
32. Chatterjee, K.; Zavadskas, E.K.; Tamošaitienė, J.; Adhikary, K.; Kar, S. A Hybrid MCDM Technique for Risk Management in Construction Projects. *Symmetry* **2018**, *10*, 46. [CrossRef]
33. Yang, J.; Tang, Z.; Jiao, T.; Muhammad, A.M. Combining AHP and genetic algorithms approaches to modify DRASTIC model to assess groundwater vulnerability: A case study from Jianghan Plain, China. *Environ. Earth Sci.* **2017**, *76*, 1–16. [CrossRef]
34. Javanbarg, M.B.; Scawthorn, C.; Kiyono, J.; Shahbodaghkhan, B. Fuzzy AHP-based multicriteria decision making systems using particle swarm optimization. *Expert Syst. Appl.* **2012**, *39*, 960–966. [CrossRef]
35. Aller, L. *DRASTIC: A Standardized System for Evaluating Ground Water Pollution Potential Using Hydrogeologic Settings*; Robert, S.K. , Eds.; U.S. Environmental Protection Agency: Washington, DC, USA, 1985.
36. Roszkowska, E. Multi-criteria Decision Making Models by Applying the Topsis Method to Crisp and Interval Data. In *Multiple Criteria Decision Making / University of Economics in Katowice*; Publisher of The University of Economics in Katowice: Katowice, Poland, 2011; pp. 200–230. Available online: https://mcdm.ue.katowice.pl/files/papers/mcdm11(6)_11.pdf (accessed on 5 September 2019).
37. Zavadskas, E.; Zakarevičius, A.; Antucheviciene, J. Evaluation of Ranking Accuracy in Multi-Criteria Decisions. *Inform. Lith. Acad. Sci.* **2006**, *17*, 601–618. [CrossRef]
38. Srikrishna, S.; Reddy, S.; Vani, S. A New Car Selection in the Market using TOPSIS Technique. *Int. J. Eng. Res. Gen. Sci.* **2014**, *2*, 177–181.
39. Data, N.O. Open Data for All New Yorkers. Available online: https://opendata.cityofnewyork.us/ (accessed on 5 September 2019).
40. Group, W.R. 2010–2013 New York City Traffic Estimates. Available online: https://lab-work.github.io/data/ (accessed on 5 September 2019).
41. Herrera, J.C.; Work, D.B.; Herring, R.; Ban, X.J.; Jacobson, Q.; Bayen, A.M. Evaluation of traffic data obtained via GPS-enabled mobile phones: The Mobile Century field experiment. *Transp. Res. Part C Emerg. Technol.* **2010**, *18*, 568–583. [CrossRef]
42. Yang, D.; Zhang, D.; Zheng, V.W.; Yu, Z. Modeling User Activity Preference by Leveraging User Spatial Temporal Characteristics in LBSNs. *IEEE Trans. Syst. Man Cybern. Syst.* **2015**, *45*, 129–142. [CrossRef]
43. Herdiansyah, S.; Sugiyanto; Guntur Octavianto, A.; Aritonang, E.; Nova Imaduddin, M.; Dedi; Rilaningrum, M. Capacity Analysis Of Parking Lot And Volume Of Vehicle Toward Sustainable Parking Convenience. *IOP Conf. Ser. Earth Environ. Sci.* **2017**, *88*, 012031. [CrossRef]
44. NYC Department of City Planning. Manhattan Core: Public Parking Study. Available online: https://www1.nyc.gov/assets/planning/download/pdf/plans/manhattan-core-public-parking/mncore_study.pdf (accessed on 25 August 2020).
45. FourSquare. The Trusted Location Data and Intelligence Company. Available online: https://foursquare.com/ (accessed on 25 August 2020).
46. FourSquare. Create Magical Real-World Moments for Your Users. Available online: https://developer.foursquare.com/ (accessed on 26 August 2020).
47. Footprints, N.B. Shapefile of Footprint Outlines of Buildings in New York City. Available online: https://data.cityofnewyork.us/Housing-Development/Building-Footprints/nqwf-w8eh (accessed on 26 August 2020).

Article

Exploring Hybrid-Multimodal Routing to Improve User Experience in Urban Trips

Diego O. Rodrigues [1,2,*], Guilherme Maia [3], Torsten Braun [2], Antonio A. F. Loureiro [3], Maycon L. M. Peixoto [1,4] and Leandro A. Villas [1]

1 Institute of Computing, University of Campinas, Campinas 13083-852, Brazil; mayconleo@lrc.ic.unicamp.br or maycon.leone@ufba.br (M.L.M.P.); leandro@ic.unicamp.br (L.A.V.)
2 Institute of Computer Science, University of Bern, 3012 Bern, Switzerland; torsten.braun@inf.unibe.ch
3 Department of Computer Science, Federal University of Minas Gerais, Belo Horizonte 31270-901, Brazil; jgmm@dcc.ufmg.br (G.M.); loureiro@dcc.ufmg.br (A.A.F.L.)
4 Department of Computer Science, Federal University of Bahia, Salvador 40170-110, Brazil
* Correspondence: diego@lrc.ic.unicamp.br or diego.oliveira@inf.unibe.ch

Abstract: Millions of individuals rely on urban transportation every day to travel inside cities. However, it is not clear how route parameters (e.g., traffic conditions, waiting times) influence users when selecting a particular route option for their trips. These parameters play an important role in route recommendation systems, and most of the currently available applications omit them. This work introduces a new hybrid-multimodal routing algorithm that evaluates different routes that combine different transportation modes. Hybrid-multimodal routes are route options that might consist of more than one transportation mode. The motivation to use different transportation modes is to avoid unpleasant trip segments (e.g., traffic jams, long walks) by switching to another mode. We show that the possibility of planning a trip with different transportation modes can lead to improvement of cost, duration, and quality of experience urban trips. We outline the main research contributions of this work, as (i) an user experience model that considers time, price, active transportation (i.e., non-motorized transport) acceptability, and traffic conditions to evaluate the hybrid routes; and, (ii) a flow clustering technique to identify relevant mobility flows in low-sampled datasets for reducing the data volume and allow the execution of the analytical evaluation. (i) uses a Discrete Choice Analyses framework to model different variables and estimate a value for user experience in the trip. (ii) is a methodology to aggregate mobility flows by using Spatio-temporal Clustering and identify the most relevant of these flows using Curvature Analysis. We evaluate the proposed hybrid-multimodal routing algorithm with data from the Green and Yellow Taxis of New York, Citi Bike NYC data, and other publicly available datasets; and, different APIs, such as Uber and Google Directions. The results reveal that selecting hybrid routes can benefit passengers by saving time or reducing costs, and sometimes both, when compared to routes using a single transportation mode.

Keywords: big data; flow clustering; intelligent transportation systems; multi-source data analyses; spatio-temporal data analyses; user experience

Citation: Rodrigues, D.O.; Maia, G.; Braun, T.; Loureiro, A.A.F.; Peixoto, M.L.M.; Villas, L.A. Exploring Hybrid-Multimodal Routing to Improve User Experience in Urban Trips. *Appl. Sci.* **2021**, *11*, 4523. https://doi.org/10.3390/app11104523

Academic Editor: Leon Rothkrantz

Received: 9 April 2021
Accepted: 9 May 2021
Published: 15 May 2021

1. Introduction

Intelligent technologies allow industries and governments to serve citizens better through disruptive applications. Among them, we highlight the ones that are related to Intelligent Transportation Systems (ITS), which are responsible for ensuring efficient and sustainable transportation [1]. Many of these applications explore data from several sources to better understand the city dynamics and adapt to them. This constant search for improvement in urban transportation stimulates new ideas and techniques to help city planners and public administrators better understand urban mobility dynamics.

The study of human mobility in urban scenarios has gained more attention recently due to the popularization of location-tracking systems [2], such as Global Positioning Systems (GPS), cellphones, and interactions in location-based social networks. These tracking solutions allow a better understanding of people's movement in a city using buses, subways, taxis, and other means. This understanding is obtained by analyzing their digital footprints. Some smart mobility solutions use data that were collected from human mobility to propose, for instance, urban routing, which focuses on achieving different objectives while suggesting routes to perform trips in the city.

Most of the current urban routing applications focus on the shortest or fastest route identification problem; however, different aspects may play essential roles in route selection. In this sense, several studies have investigated different aspects of urban trips. For instance, routing approaches focused on non-motorized means of transport, also called active transportation (e.g., walking, bicycle) [3], enhancing the levels of enjoyment in the trip [4], or even identifying users' preferences using probabilistic frameworks [5].

This paper investigates the use of different transportation modes as a substitute to traditional modes. We devise a model that allows the evaluation of transportation systems in cities with various transportation modes. The same model can be used to recommend routes by also applying user preferences. We describe both cases in this work. Our solution creates hybrid-multimodal routes that combine different transportation modes on a trip. This approach differs from traditional multimodal routing, which suggests multiple options, each with a single transportation mode. We use historical trip data from taxis of New York City to evaluate the hybrid-multimodal routes. The main objective of this study is to evaluate how a different mechanism for creating and recommending routes for urban trips can impact the user. Our model's main contribution is the recommendation of urban trips that is aware of not only cost and duration, but that also is subjective to users' perceptions. Approaches to model perception of users when recommending urban routes have recently gained relevance in the context of smart cities, for instance, with eco-friendly route recommendation [6] being adopted in mainstream route recommendation apps, such as Google Maps.

This work extends our previous study [7], where we proposed a method to improve route selection through hybrid private vehicles and traffic information. Previously, the route selection was made by evaluating the duration and cost of urban trips. Here, we advance our previous work by exploring a novel user experience framework that is based on Discrete Choice Analysis [8]. This approach allows us to compare different route options while accounting for variables, such as the acceptability of active transportation modes, cost of transfer between different modes, and user preferences through a user profile operator. Besides that, we also implement other improvements, such as consider more alternative modes (e.g., bicycle) and other datasets to evaluate users' impressions of urban trip segments. The novel research contributions of this work are: (i) an origin-destination flow clustering technique that groups urban mobility flows and classifies them in trending or secondary flows; and, (ii) the proposal of a user experience model and a profile operator to evaluate urban trips. Additionally, we evaluate more transportation modes than our initial work and propose a strategy to deal with anonymized data.

The rest of this paper is organized, as follows: Section 2 discusses studies that are related to the proposed technique, i.e., clustering, modeling, and routing. Section 3 provides an overview of the methods used to evaluate our model, which comprises the proposed flow clustering methodology. Section 4 presents our model to evaluate route options for a given trip and how it can be personalized for each user. Section 5 analyzes the results. Finally, Section 6 presents the concluding remarks.

2. Related Work

We study different areas to produce the hybrid multimodal routing methodology. In particular, we propose a mobility flow clustering technique, a hybrid-multimodal routing algorithm, and a user experience model for urban trips. Below, we discuss proposals for clustering mobility data (Section 2.1); and, routing algorithms and models for route selection (Section 2.2).

2.1. Clustering

The clustering of mobility flows is a research field that creates algorithms to identify trip patterns. There are two main types of clustering in urban scenarios: (i) trajectory-based clustering: the whole trajectory of the moving entity is known, and used [9,10]; and, (ii) Points-of-Presence based clustering: only a few points that are related to the moving entity are known [11–13]. Their main difference is the sampling rate of observed points. In this work, we focus on exploring datasets with data regarding pick-up and drop-off locations only; thus, our proposed algorithm fits into the second type of clustering.

Pang et al. [11] proposed a method to cluster the pick-ups and drop-offs of taxi trips. They spread the points in a fine-grained matrix and write each cell's frequency in it, which is decomposed and factorized. The resulting values are plotted as a heatmap over the city's map. Their method identifies hotspots of trips in the city, which allows them to evaluate their correlation with public transportation stations, business areas, and airports. The structure of the outcomes is the main difference to our solution. Their method focuses only on identifying regions, and ours in identifying the regions and the flow between them, which is an important aspect in the route selection.

Stepwise Spatio-Temporal Flow Clustering (SSTFC) [14] is a spatio-temporal flow clustering technique that identifies mobility trends. SSTFC has spatial and temporal steps and it works likewise other proposed mechanisms in tge literature [15]. In the spatial step, a minimum neighborhood and a size coefficient are used with a custom distance metric to decide whether to merge or not neighboring clusters. Similar parameters exist in DBSCAN [16], namely *min_samples* and *eps*, which are also used to evaluate the density of a region and create clusters. After the spatial step, SSTFC uses the same method to create temporal clusters inside the spatial ones, but using a third parameter, a temporal threshold, to merge neighboring clusters. In [14], it is assumed that SSTFC is robust, but there is no quantified assessment. As clusters get merged several times, flows with varied directions are mixed in the same group, which is sub-optimal when trying to identify mobility trends. Our algorithm first locates the functional departure and arrival zones, which limits the flow directions. These flows can have different angles, but they always flow from one region to another, while, in their technique, flows going forward/backward might end at the same cluster.

2.2. Routing and Modeling

Different factors affect decision making when choosing urban trip routes. Most commercial systems and IT studies map the problem to the shortest/fastest pathfinding [5,17]. The literature of Discrete Choice Modeling [8] and Transport Economics [18] for route choice modeling is well-studied. Yet, some proposed models are prohibitive due to data restrictions that are present in real-world applications and the difficulty in assessing every aspect that drives route selection. Our model builds upon discrete choice modeling frameworks and defines premises to ease the implementation in the real world. Additionally, we identify data sources that can be used to better model the user's behavior.

This study assumes that cost, duration, and quality of experience (QoE) play a central role in the mode/route choice process. Other studies model these choices while considering more variables. The FAVOUR algorithm [5], for instance, uses a probabilistic model over historical data. It represents the route's cost/utility as a weighted sum of the cost of route segments. Every segment is evaluated with different features, leading to different weights for the same segment. Furthermore, the algorithm considers that some route options have costs that are associated with the entire route, rather than every segment to contemplate different phenomena, such as weather. We use similar variables to those that are used in FAVOUR, such as distance, duration, price, and the number of transport mode changes. However, the FAVOUR algorithm does not consider the acceptability of using active transportation modes, which we do. In their study, walking and cycling long distances are evaluated in the same way as short ones. Another difference is that we combine the features into a familiar metric to users, such as a monetary price or a duration. The FAVOUR model is based on a Bayesian learning technique to define the weights for the sum and recommend routes for users. Differently, we use Discrete Choice Analysis [8] to model the costs of a trip and apply nominal values for the quality of service (QoS). Finally, we use a profile operator to suggest routes separately that provide ways to compare routes with or without the bias of individuals. FAVOUR always counts on the existence of individual user data to adapt its recommendations. Finally, our model considers traffic conditions to select transition mode points, while FAVOUR does not consider real-time traffic information.

Hrnčíř et al. [19] proposed a method for selecting bicycle routes using different criteria. They defined a generic framework based on a graph with geo-coordinates, altitude, the horizontal length of cycling paths segments, and other features of route segments. For the experiment, they defined the features to be the ones that are provided by OpenStreetMap (https://www.openstreetmap.org/, accessed on 14 May 2021). The authors modeled the route selection problem as multi-criteria search optimization and proposed heuristics to make this search feasible. This study focuses on maximizing different utility functions that are applied to routes. They do not provide methods to obtain scores that aid in ranking or personalizing the results, such as our proposal. They only consider one transportation mode (i.e., bicycles), which causes important variables to be omitted, such as the acceptability to chose a given route using active transportation. When taking active transportation, users may feel discouraged to perform long trips, which is considered in our model.

3. Evaluation Framework

This work proposes a method to combine different transportation modes for single urban trips. For each trip, we evaluate its duration, price, and user experience. Section 3.1 briefly introduces the concept that we use to evaluate urban trips. Section 3.2 describes the methodology used to combine transportation modes, produce multimodal-hybrid routes, and its evaluation. This solution explores real-time traffic data to replace trip segments using taxis with other transportation modes (e.g., bicycle or public transportation). In order to evaluate the hybrid-multimodal routes, we use datasets from the city of New York, available at the NYC Open Data portal (https://opendata.cityofnewyork.us/, accessed on 14 May 2021). We performed data reduction by clustering the main mobility flows due to their size, as described in Section 3.3. When looking for route choices, we run different nearest neighborhood and temporal queries. To reduce time executing these queries, we combine two different data indexes into a single data structure creating the Multi-Layer Geographical Linkstream, as detailed in Section 3.4.

3.1. Overview

Figure 1 illustrates the process of evaluation of urban trips. We start with a dataset containing origins and destinations of urban trips. After, we use our proposed mobility flow clustering technique, as described in Section 3.3, to create a weighted graph of the urban flows. This step makes feasible the evaluation of huge datasets by applying the model on the most relevant flows. Once the graph is obtained, we evaluate possible driving ways for a given flow and combine these driving ways with traffic data to identify congested segments in the trip. These segments then are replaced with alternative transportation modes in the route combination phase. After all traditional and hybrid route options are obtained, we assess the gains/losses in terms of duration and cost, and also we use our proposed user experience model (as described in Section 4) to evaluate the quality of all options.

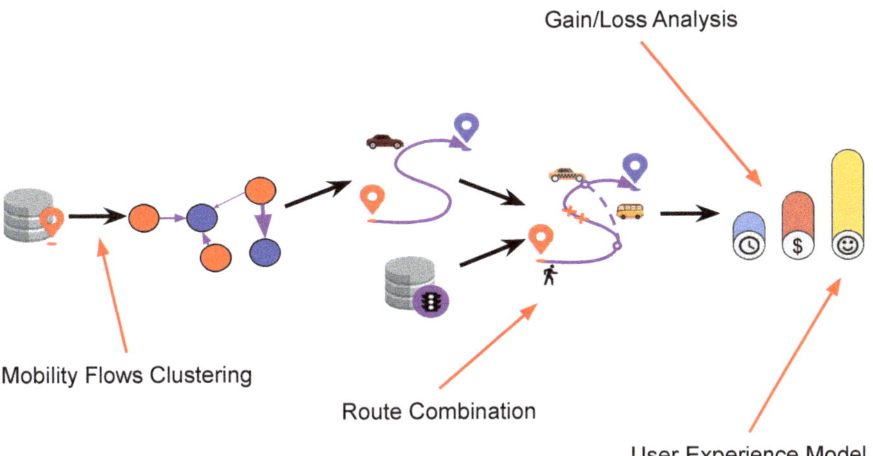

Figure 1. Overview of proposed methodology.

3.2. Methodology

We propose a Multi-Layer Geographical Linkstream (MLGLS) to organize data from different sources to assess the impacts of using multimodal hybrid routes for urban trips. We study the impacts of replacing taxi-only trips in New York with hybrid options using our proposed methodology, such as reducing time spent on traffic jams, which leads to an increase in QoE during the trip. These hybrid routes contrast to the traditional approach of multimodal routing, which consists of different options using single modes (e.g., one public transportation, one driving, and one walking option). The hybrid approach that was proposed in this work also considers routes that started with one of the traditional modes and finished in another (e.g., started in public transportation and finished in a taxi). The purpose of this section is to describe the methodology used to evaluate the hybrid routes. The process is divided into seven steps: (i) reduce data about trips from Yellow and Green Taxis datasets and find main mobility flows; (ii) create and populate an MLGLS with trip data and with other transportation modes (e.g., bus stations, bike dockers); (iii) evaluate the initial driving options that contain traffic information; (iv) identify congested segments in the initial options; (v) select mode transition points and combine them; (vi) trace route alternatives and save them; and, (vii) evaluate the metrics of the alternatives. Steps i, ii, and vii are executed once for the whole dataset. In contrast, the other steps are repeated to every mobility flow in Step i. For each one of these flows, the trip is evaluated when considering the spatio-temporal ⟨origin, destination⟩ pair. Additionally, changes in the scenario during the trip are not accounted for. Figure 2 outlines the methodology.

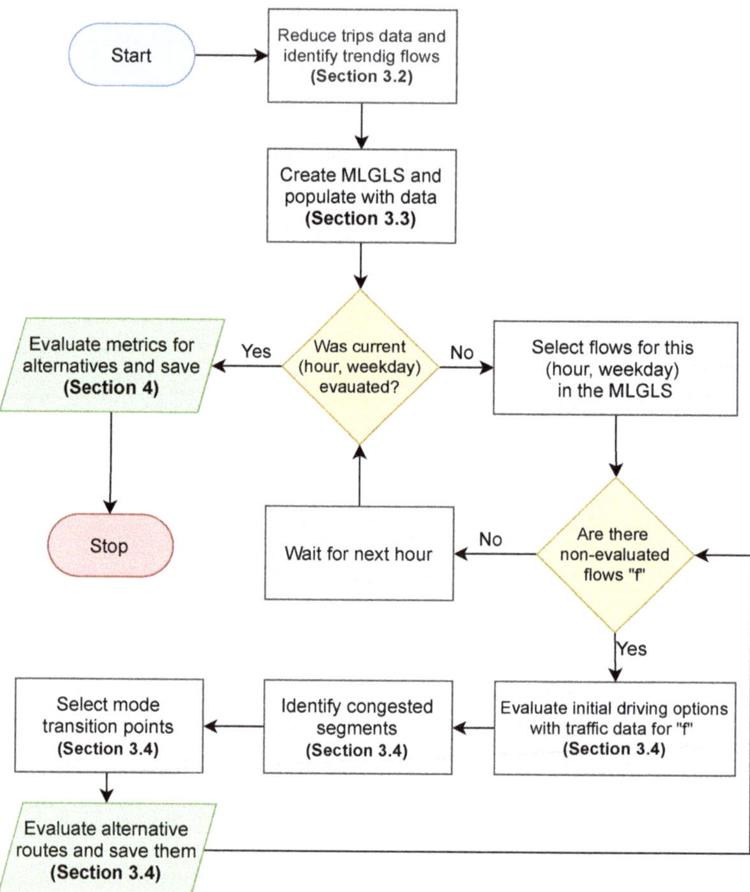

Figure 2. Work flow to evaluate the impact of using hybrid-multimodal routes to replace taxi trips in New York.

Section 3.3 describes Step i, Section 3.4 give details about the MLGLS used in Step ii. Section 3.5 comprises Steps iii. iv, v, and vi. Section 5 explains the outcomes that are produced in Step vii, and other important intermediary results, such as the individual route selection evaluation using the model described in Section 4.

3.3. Data Reduction

The proposed model analyzes trip-by-trip, every route option by requesting traffic data, trip fares, and route duration estimates through different APIs in real-time over the internet to assess the price, duration, lengths, and waiting times of trip segments. This analytic approach demands a long time to perform data collection and analysis. Thus, to have a feasible study case, we have a data reduction step that clusters mobility flows. This step aims to identify the city's main flows to perform the evaluation. The data reduction technique that is described below is general enough to be used in different datasets and for different purposes.

The technique identifies trip Origin-Destination Mobility Flow Corridors (OD-MFC). This is accomplished by finding functional zones on the coordinates of the dataset and then computing the flows between these zones. It has two variants to adapt to different aspects of datasets. If the dataset has geographical coordinates (i.e., latitude and longitude), the full version of the technique must be applied. However, there are plenty of datasets that are anonymized, only containing region ids where the departures and arrivals happened. For these datasets, a light version of the technique was developed.

Both of the versions of the OD-MFC technique have three steps: (i) Split (pre-processing): divides the dataset into chunks by considering a division between hours, weekdays, date or other variants—we used the division ⟨hour, weekday/weekend⟩, e.g., ⟨19 h, weekend⟩, which describes the class with trips between 19 h and 20 h on Saturday and Sunday; (ii) Functional Region Identification: clusters the departures and arrivals of trips to identify the most relevant zones; (iii) Flow Accounting: is responsible for accounting the flows between those identified zones; and, (iv) Flow Classification: classifies the flows in trending (i.e., flows with a more relevant amount of trips) and secondary.

The pre-processing step divides the dataset into smaller ones that capture the data seasonality. In our study, we defined this division by analyzing different shapes of distribution curves of the trips. Afterwards, the second step identifies the functional regions in the data (i.e., the most relevant zones). This is the single step in which the full and the light versions of the technique differ. For the full version, we use the HDBSCAN [20] clustering algorithm to identify the functional zones. This algorithm is widely adopted to cluster positional datasets containing noise. For the light version, the regions are already identified. Thus, all of the regions could be used in the analysis.

The third step consists of accounting for the flows between regions. Each flow is described by the number of trips between two regions. Finally, the fourth step identifies the main flows. The classification technique was inspired by a similar strategy that was proposed to identify the EPS in the DBSCAN algorithm [16]. In their paper, the authors evaluate the k-nearest neighbor distance for all samples in the dataset and plot them sorted. This plot tends to form an exponential decay curve. Based on this technique, we created a that sorts the flows based on their magnitude and identifies the "knee" of the resulting curve. We use Equation (1), the Curvature for a Plane Curve [21] equation, to identify this "knee".

$$\kappa = \frac{|y''|}{(1+y'^2)^{3/2}}. \tag{1}$$

The function y is the curve that formed by the sorted flow magnitudes. This equation evaluates the magnitude of the curvature at a certain point in the curve. Because we are looking for the "knee" of the curvature, we must look for local maximum for κ, i.e., $\kappa' = 0$. A previously faced problem of this study [7] was that the curvature formed by real data was not a smooth curve. Thus, to enhance calculating the required derivatives, we now smooth the data, depending on the desired results, before applying the classifier. In this work, we use interpolation with an exponential function to smooth our data. We select the "knee" point in the curve, which is used as a threshold to classify the flows in trending or secondary. If we use an exponential function of the form $y = e^{ax}$ to model the exponential decay, then the solution of Equation (1) leads to:

$$\kappa = \frac{a^2 e^{ax}}{(1+a^2 e^{2ax})^{3/2}}, \tag{2}$$

where the maximum is given at:

$$\kappa' = 0 \longrightarrow x = -\frac{\log 2a^2}{2a}. \tag{3}$$

In this study, we use HDBSCAN algorithm [20] to identify functional regions. HDBSCAN is a $O(n^2)$ time algorithm, with approximate solutions being given in average $n \log n$ time [22]. The proposed classification approach has to first sort mobility flow popularity in time $O(m \log m)$, with $m \leq n$, but, in general, $m << n$. After that, the approach needs to solve a minimization problem to identify the best fit exponential equation to fit the curve that formed by the sorted popularities. We use the Levenberg-Marquardt method, which takes $O(\epsilon^{-2})$ time [23]—ϵ is the precision adopted. After identifying the best fit curve, the classification is done in $O(1)$. Therefore, the approach is executable in $O(p^2)$, where $p = \max\{n, 1/\epsilon\}$. The proposed method maintains the same time complexity of HDBSCAN.

To assure the quality of the proposed clustering technique, we first verify whether it complies with the requirements of good clustering techniques [24], i.e., the clustering: (i) should not impose a priori bias on the clusters' shape; (ii) should be able to handle varying densities; and, (iii) should be able to handle varying dimensionality. The origin and destination functional regions are obtained using the HDBSCAN technique, which does not impose a bias on the clusters' shape, and it is also known for good performance with noisy datasets. Additionally, the selection of the trend/secondary flow threshold identifies a point with no bias, thus meeting requirements (i) and (ii) (as discussed in Section 5.3). Regarding requirement (iii), despite using a spatial 2D dataset in this study, neither HDBSCAN nor the curvature equation limits the number of dimensions of the dataset. Thus, the proposed technique can handle higher dimensionalities. We conduced a experiment comparing our approach to the SSTFC algorithm [14]. Section 5 shows that both versions of OD-MFC outperform SSTFC when identifying the most relevant flows in noisy datasets.

3.4. Data Structure

We model all the data in MLGLS data structure, which was created by combining: (i) Linkstreams [25], a temporal graph structure that stores links sorted according to time of occurrence; (ii) KDTree [26], which is a tree structure to ease nearest neighbor searches; and, (iii) Multi-Aspect Graph [27], a derivation of static graphs that allow the creation of multiple layers to represent data features.

A link is a 3-tuple (a, b, t) that shows a contact between a and b at time t. Contacts that are represented in this stream are time-sorted and they may optionally have a duration, in which case the Linkstream may be called a Stream Graph. Because links are sorted in time, one can use a binary search to optimize queries. However, the Linkstreams are not designed to deal with positions, an important feature of mobility data. To tackle this issue, we added a complementary index to the Linkstream using KDTrees. The way that this tree is constructed allows the quick execution of neighborhood searches, as opposed to other spatial trees, such as the R*-tree [28].

The use of Linkstream and KDTree makes it possible to run quick spatio-temporal data searches. However, urban mobility is heterogeneous and it can be collected from different sources, and must be combined to perform a more insightful analysis [29]. Such a process must not lose information about the data origin since data from different contexts may have different interpretations. To combine data without losing its source information, we use a feature from the Multi-Aspect Graphs, the possibility of creating multiple layers in the graph to represent different data sources. This data structure allows for the creation of aspects, a data dimension where multiple layers can be added. Hence, we propose the usage of several layers where data from different sources can be sheltered.

Formally, the MLGLS is a group of nodes $v \in V$, connected by links $e \in E$. These nodes and links may hold further information detailing the event identified by the node or the type of contact that is indicated by the edge. The existence of events (or nodes) are known via a data source (or layer) $l \in L$; and, all of the interactions happen in a time interval $T = [t_0, t_f]$. Additionally, to enhance the spatial queries' performance, we have a Geographical Index I (i.e., the KDTree). Thus, the MLGLS is given by the tuple $M = \langle V, E, L, T, I \rangle$.

An important implementation aspect is the existence of different layer types. The data structure supports both static graphs and linkstream layers. For static layers, the difference is that nodes and links exist during the whole time interval, from t_0 to t_f. Because of the characteristics of modern datasets, there is another type of layer, called dynamic. This type is used for layers in which nodes and edges vary in time, but the information regarding these graph entities can only be obtained when needed. For example, the bike dockers in our experiments may or may not have bicycles available at a given moment. Thus, when we need to verify the existence of a node, we perform a call to the provider API to evaluate if that node will, or will not, be listed in the graph. The dynamic layer is defined in the MLGLS as a time-dependent function called upon the necessity to evaluate the current state of a layer.

Figure 3 depicts our proposed data structure, which has two dimensions: time and data sources. Layers can be created at each dimension, being represented by rows or columns of rectangles in Figure 3. The dotted blue line L_h is surrounding the taxi data layer in the data-source dimension, and the green dotted line L_v is surrounding the time layer t_2. The nodes (circles) are the data entries, and each link (continuous line) represents the contact between those entries. The dashed thick lines represent contacts that persisted in time. Finally, the red lines represent the spatial index that is created by the KDTrees. This index creates perpendicular divisions in the space to separate nodes and ease queries.

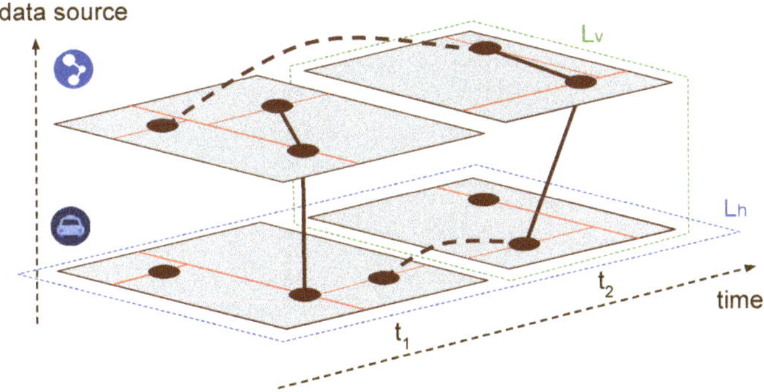

Figure 3. Example of Multi-Layer Geographical Linkstream.

When merging transportation modes to create alternative urban routes, we fuse five positional datasets: subway stations, bus stops, bicycle dockers, and the Yellow and Green Taxis trips. When analyzing these datasets, we have to query data in time and space; the proposed data structure saves time and resources. Additionally, it is general enough for usage in other scenarios. The approach used to combine these datasets and produce hybrid-multimodal route alternatives is described below.

3.5. Route Combination

The processes between Steps iii and vi (Section 3.2) are repeated for every flow inside its respective ⟨hour, weekday⟩ class. A flow is defined by an origin, a destination, and a weight based on the number of trips. For every flow, we obtain a list of alternative routes, including the traditional, and the hybrid options to be compared. Algorithm 1 [7] shows how this list of alternatives is produced.

Algorithm 1 Evaluate route alternatives for trip flows.

```
    Input: Origin and destination of a flow.
    Output: List with the route alternatives for the flow.
 1: drive_way ← get_driving_way(origin, destination)
 2: start_transitions ← new List()
 3: end_transitions ← new List()
 4: options ← new List()
 5: previous_step ← none
 6:
 7: push(start_transitions, origin)
 8: push(end_transitions, destination)
 9:
10: for step in drive_way.steps do
11:    if is_congested(step) then
12:       if is_congested(previous_step) then
13:          pop(end_transitions)
14:          push(end_transitions, step.destination)
15:       else
16:          push(start_transitions, step.origin)
17:          push(end_transitions, step.destination)
18:       end if
19:       previous_step ← step
20:    end if
21: end for
22:
23: for st in start_transitions do
24:    for et in end_transitions do
25:       opts ← get_options(origin, st, et, destination)
26:       concat(options, opts)
27:    end for
28: end for
```

Algorithm 1 starts (Line 1) by calling the `get_driving_way` function (Step iii), with the origin and destination as arguments. This function calls the TomTom Routing API (https://developer.tomtom.com/, accessed on 14 May 2021), which returns a list of steps to perform the trip and traffic data. In Lines 7 and 8, the algorithm appends the origin and destination to the list of transition starts and ends. They are added to create the possibility of full trips being performed in single modes (i.e., the traditional routes). The `foreach` command in Line 10 populates the transition lists according to the state of traffic (Step iv). This state is evaluated via the function `is_congested`, which indicates whether it has traffic or not. This binary classification is only used to produce the route options. When evaluating the Quality of Experience (QoE), we consider the time that would be spent on the congestion, which is assessed based on historical data using the TomTom API. Note that the steps with bad traffic conditions in a row are merged in the **if** command in Line 12 just by updating the `end_transitions` list.

Lines 23 and 24 contain the `foreaches` that are responsible for combining the transition points (Step v). Inside these `foreaches`, a call to the `get_options` function evaluates all alternatives considering a tuple with an origin, destination, start transition, and end transition. This function uses the data populated in the MLGLS to identify the actual transition points, when considering mode characteristics (e.g., bus stops). Next, the function evaluates the possible options and saves data about traffic, waiting times, duration, and prices for every step and the mode used to compose it.

The resulting list of options is concatenated to a list with all options (Line 26) to be saved, as described in Step vi. After evaluating all of the route alternatives, they are compared with the traditional routes using the metrics duration, price, and user experience (Step vii). Section 4 describes how to evaluate the user-experience model. The route combination and analysis code is open source, being distributed under the SMAFramework [30] on Github (https://github.com/diegopso/smaframework, accessed on 12 May 2021).

4. User Experience Model

This section describes the evaluation of user experience in urban routes, which is based on the Opportunity Value [31]. This value is a technique used in Discrete Choice Analysis to assess the utility of a decision made in a set of mutually exclusive options. This decision is the same when picking a route in an urban trip: once a route is chosen, the others become unavailable. Sections 4.1–4.3 model the different deterministic aspects of the route. The models in these sections can be used to evaluate the transportation system conditions and various transportation modes. Section 4.4 presents the experiment instances that are used to evaluate user satisfaction. Section 4.5 introduces a Profile Operator used to quantify the impact of personal choices in the model. It works on top of the deterministic evaluation and adds a probabilistic dimension regarding individual preferences and, thus, enables the model to be used when performing route recommendations.

4.1. Core

Let $r \in R$ be a route option to a given trip of a person (decision-maker). R is the set of available options created with a heuristic. r is a tuple $\langle t_r, c_r, l_r, S_r \rangle$, the route's duration, cost, length, and set segments in different transport modes, respectively. One way of comparing aspects of route options of a trip is evaluating the opportunity cost of these aspects. For instance, the Opportunity Cost Equation [31] (Equation (4)) evaluates the opportunity cost for selecting a route in the set R as the loss due to selecting this option and not the best possible option.

$$copp_t(r) = t_r - \min_{\forall r' \in R} t_{r'}. \tag{4}$$

where r and r' are the routes taken from the set R with duration t_r. $copp_t(r)$ denotes the opportunity cost of duration of a given route r. The opportunity costs of other route aspects are given by analog equations evaluated on c_r, l_r, and so on.

A typical use of the opportunity cost in transport engineering is to evaluate the Value of Time Travel Savings (VTTS). It represents the monetary cost of the extra time that is spent by selecting a given route instead of the fastest route.

Nevertheless, in order to obtain the user experience in the trip, we must consider both the value of the user's time and the actual paid value, instead of only the duration of t_r. We can use the Generalized Cost Equation [32] (Equation (5)), also used in transport economics when evaluating the attractiveness of a route option [33], to combine the two variables. The generalized cost of a route option is the addition of the paid monetary cost to the generalized journey time, which was found via a utility function.

$$g(r) = c_r + u(r). \tag{5}$$

where $g(r)$ is the generalized cost, c_r the monetary cost of the route, and $u(r)$ the utility function to estimate a generalized time. The generalized time u is estimated using the Value of Time Travel (VTT) of the user. u varies with the context of the decision-maker, such as route conditions and his income/hour. One way of evaluating u is by multiplying the trip duration by the VTT of the decision-maker, as $u(r) = \tau t_r$, where τ is given in [USD/h]. This may be used to compare the routes with similar characteristics. However, when comparing multiple transportation modes, the impressions of the decision-maker might vary among different modes. One way of contemplating this phenomenon is by using a Generalized Time Function [34] that considers the user perceptions to adapt the measured time. For instance:

$$u(r) = \tau T(r). \tag{6}$$

Subsequently, to define the general time of the complete route $T(r)$, we sum the general time of all segments $s \in S_r$:

$$T(r) = \sum_{\forall s \in S_r} T(s), \tag{7}$$

where $T(r)$ is the generalized time in the complete route, and $s \in S_r$ is a route segment in a specific transportation mode, which has a duration, cost, length, and its mode $s = \langle t_s, c_s, l_s, m_s \rangle$.

4.2. Mode Transfer Cost

When a user changes transportation mode, he incurs different costs. For instance, when changing to a private hired vehicle, companies usually practice a hiring fee that increases the monetary cost of the trip. Furthermore, some transportation modes require a waiting time, until a vehicle arrives to take the user, which increases the total duration of the trip. Hiring fees and waiting time for private hired vehicles and busses are both assessed in our experiments; they are obtained from real data through different APIs, e.g., Uber and Google Directions. Besides these costs, we use a mode transfer penalty to model the direct costs in terms of experience that the user incur by having to change the transportation mode. We introduce an exponential depreciation operator in Equation (8) to account for mode transfer penalty cost in experience for the user. We count the number n_r of changes in the route and use it to compute the total impact in a given route option as:

$$\lambda_r = \begin{cases} 0, \text{ if } n_r = 0 \\ e^{a n_r}/100, \text{ otherwise} \end{cases} . \tag{8}$$

where λ_r is the impact due to the number of mode changes. In our study, we did not obtain supporting data to evaluate a, this parameter could vary according to user preferences. Yet, $a = 1$ results in a reasonable impact considering traditional urban trips standards. Table 1 shows the impact according to the number of mode changes for $a = 1$.

Table 1. Transfer mode impact for $a = 1$.

n_r	0	1	2	3	4	5
λ_r [%]	0%	2.72%	7.39%	20.09%	54.60%	148.41%

Using Equation (8), the final generalized time of a route in Equation (7) changes to Equation (9).

$$T(r) = (1 + \lambda_r) \sum_{\forall s \in S_r} T(s). \tag{9}$$

4.3. Experience in Different Route Segments

The satisfaction of the decision-maker in a trip segment varies according to segment accessibility. Thus, to measure this satisfaction, the general time for a given segment is:

$$T(s) = S(s) t_s + \epsilon, \tag{10}$$

where $S(s)$ measures user satisfaction in a given transportation mode. The variable ϵ describes the error due to non-observable components, such as weather conditions. Here, we consider that this error is normally distributed, which causes a reduced impact on the final values according to the Random Utility Theory [18].

To define S, we use In-Vehicle Time (IVT) multipliers that estimate impressions of a user in a given transport mode as compared to in-vehicle impressions. For example, a given user may consider 5 min of walking to be as unpleasant as 10 min in a vehicle. In this case, the IVT multiplier would be $10/5 = 2$. There are different ways of estimating IVT values, which usually consider surveying passengers. We consider the characteristics of the segment s and impressions of the level of service of the decision-maker w to measure the satisfaction S.

$$S(s) = w(s)P(s). \tag{11}$$

$P(s) \in [0, 1]$ is a depreciation factor to describe the acceptability of traveling a segment s. Functions P and w depend on the scenario, and their selection is discussed in Section 4.4.

4.4. Experiment Instances

To evaluate the satisfaction of users in route segments in the present study, we consider that both functions w and P vary with the transportation mode. Additionally, the impression about the level of service in a segment depends on the time spent; and, the acceptability of a given segment is a function that depends on the length of the segment, i.e.,

$$S(s) = w_{m_s}(t_s)P_{m_s}(l_s). \tag{12}$$

where m_s is the mode used on the segment s; w_{m_s} and P_{m_s} are respectively the impression of the user and the depreciation function in relation to the transport mode used in that segment. We consider a limited set of modes and use data from a survey [35] to obtain Equation (13), an instance of w. We combine bicycle and walking modes, since both require physical effort to perform, i.e., active transportation. We do not have the actual multiplier for the bicycle mode, but a study conducted in Beijing suggests bike satisfaction to be high when compared to vehicles [36].

$$w_{m_s}(t_s) = \begin{cases} 1.54t_s, & \text{if } m_s \text{ is in-vehicle congestion} \\ 0.78t_s, & \text{if } m_s \text{ is in-vehicle headway} \\ 1.70t_s, & \text{if } m_s \text{ is waiting} \\ 1.65t_s, & \text{if } m_s \text{ is walking or bicycle} \end{cases} . \tag{13}$$

where the impressions of the user w_{m_s} in relation to the transport mode m_s is obtained in function of the duration of the trip segment t_s.

$P_{m_s}(l_s)$ presented in Equation (11) is a depreciation factor that decreases the user satisfaction in a segment according to its length, which depends on the transportation mode m_s. In the urban context, vehicles and public transportation usually perform the longest journeys (modes as helicopters are omitted). Thus, the impact of $P_{m_s}(l_s)$ in such routes is minimal. However, the impact on the walk and bicycle modes is highly relevant and they are usually mentioned as walkability [37] and bikeability [38], respectively. They may consider more factors other than the distance (e.g., accessibility, air pollution).

One way of guessing the likelihood of someone adopting a segment of active transportation is to observe a dataset of trips with their lengths. We can estimate the probability of choosing a segment by calculating an inverse cumulative distribution function of trip lengths in the dataset. This approach is already used in the literature to estimate walking and bicycling acceptabilities (The Netherlands [39] and US [40,41]). Moreover, these abilities vary according to the trip purpose. For everyday commute trips using public transportation, they also depend on the accessed mode—one might walk twice the distance for using rapid modes as metro or tram [41].

In this work, we estimate bikeability using data from the Citi Bike NYC (https://www.citibikenyc.com/, accessed on 14 May 2021) in New York. However, we did not have access to a dataset to estimate walkability. Thus, we use previous results that were obtained for US cities [40]. We defined both functions by interpolating the results for walkability and bikeability. P_w (Equation (14)) is used to evaluate the acceptability when walking, and P_b (Equation (15)) when using a bicycle. We also account for users' will to walk further for accessing rapid transportation modes [41]. The value of l_s for the equations is given in kilometers. Additionally, the values of P are capped between 1 and 0.01 to prevent invalid outcomes. Section 5.5 discusses walkability and bikeability, and shows the shape of the inverse cumulative distributions.

$$P_w(x) = -6.829x^4 + 9.788x^3 - 2.847x^2 - 1.858x + 1, \tag{14}$$

$$\begin{aligned} P_b(x) = & -2.11 \times 10^{-4}x^5 + 3.93 \times 10^{-3}x^4 \\ & -2.63 \times 10^{-2}x^3 + 7.50 \times 10^{-2}x^2 \\ & -8.05 \times 10^{-2}x + 1.01. \end{aligned} \tag{15}$$

4.5. Individual Route Selection

This section proposes a strategy for selecting routes among different options based on scores and they can be used to create a recommendation set. The recommendation should also consider other issues, such as the impact of recommended routes in the overall system when the adoption rate is high. The recommendation approach is responsible for mitigating this issue by, for instance, load balancing the set of reasonable routes [42]. We do not address the issue of load balancing recommendations in the present study.

The described model has three main results: (i) the total opportunity cost (Equation (4)) to be used to select a route among different choices; (ii) the generalized cost (Equation (5)), which measures an overall cost/benefit of an option considering its cost, users' perception and duration; and, (iii) the generalized time (Equation (9)), which removes the cost and only evaluates the users' perceptions of time. These metrics model general users in specific populations. However, when it comes to individual choices, these values may vary.

We introduce a profile operator to tackle the issue of different opinions of users when selecting routes. This operator is inspired by the concept of Mixed Strategies from Game Theory, which is a probability distribution over a set of actions taken by an agent [43]. We consider that the decision-maker has two Pure Strategies: always choose the shortest duration or always choose the cheapest trip.

To produce a Mixed Strategy, we allow the decision-maker to pick a probability p in which the QoE is favored, despite the cost. The remaining odds $(1-p)$ lie in the event of choosing a trip based on the cost. The Expected Utility [44] (Equation (16)) of the outcome of selecting routes, based on the Mixed Strategy that was adopted by the decision-maker, is obtained in this case, for two variables, as:

$$\mathbb{E}[P] = u(A)p + u(B)(1-p), \tag{16}$$

where p is the probability that defines the profile P and $u(A)$ and $u(B)$ are the utilities of the choices A and B, respectively. The difference in our operator is that, in our equation, we use two utility functions for the same action, resulting in Equation (17).

$$\zeta = u_1(r)p + u_2(r)(1-p), \tag{17}$$

where ζ denotes our proposed operator, $u_1(r)$ and $u_2(r)$ are the two different utility functions for the action of selecting the route r, and p, $(1-p)$ the respective complementary probabilities. If we select $u_1 = (t+T)/2$, and $u_2 = c$ from the utility functions previously described, then the profile operator ζ of a route r given the profile p becomes, as shown in Equation (18). u_1, represents the QoE-related aspects and u_2 the monetary costs.

$$\zeta(r,p) = p\left[\frac{t_r + T(r)}{2}\right] + (1-p)c_r. \tag{18}$$

where the profile operator ζ for a given route r and profile p is given in terms of the duration t_r, the generalized time $T(r)$, and the cost c_r, and its related probabilities p and $1-p$.

In Equation (18), $p \in [0,1]$ represents the value that a specific user gives to QoE-related variables. QoE represents the duration and overall user impressions about the duration.

We oppose the cost to the quality metrics, i.e., users expecting better quality might pay more for the trip, while a user expecting to pay less may have a worse service. The best route, according to the model, obtains the smallest value for ζ. To apply Equation (18), t_r, c_r and $T(r)$ must be normalized. We use the z-score normalization function for the experiments in this work. We selected the z-score, since it assesses the distance between a route aspect measurement and the average of all measurements of that same aspect for a trip. This evaluation of distance is given in terms of standard deviations; it counts how many times an option is worse than others in terms of standard deviations.

With this operator, our model is complete and it complies with Discrete Choice Modeling framework [18], which is comprised of four elements: (i) the decision-maker: the person that aims to choose a route; (ii) the alternatives: the routes; (iii) the attributes: the utility functions; and, (iv) the decision rule: the profile operator.

5. Results and Discussion

In this section, we discuss the obtained results. We start describing the different data sources used to evaluate hybrid trips in the city of New York in Section 5.1. Afterwards, we discuss the mobility flow clustering technique proposed; the pre-processing results are listed in Section 5.2, the final output of the clustering is presented in Section 5.3, and, in Section 5.4, we compare our proposal with a study from the literature. Finally, we present the results that are related to the hybrid routes and the user experience model. Section 5.5 discusses the acceptability of using active transportation modes, Sectoion 5.6 discusses the overall values for cost, duration, and user experience for the evaluated trips, and Section 5.7 shows the results for the personalizing routes based on user profiles.

5.1. Data Characterization

In this section, we briefly explain the datasets that are used to obtain the results in this study.

5.1.1. Taxi Data

The main datasets used in this work are the Yellow and Green Taxis datasets from New York (https://www1.nyc.gov/, accessed on 14 May 2021), which were taken from the NYC Open Data portal, from 2016 and 2017. Part of the period is anonymized, i.e., they do not contain geographical coordinates of the trips' pickups and drop-offs, but, instead, the IDs of regions where they started/ended. These are the two main taxi services in New York: the Yellow Taxis cover mainly the center of Manhattan Island and airports, and the Green Taxis usually cover the periphery of the city. The Yellow and Green Taxis datasets have 113.4 and 11.7 million trips, respectively. We removed invalid data from these datasets, such as trips starting and ending at the same point, vehicles traveling over 100 km/h, or trips over 6 h. The remaining trips were clustered into flows and evaluated.

5.1.2. Traffic Data

We used TomTom Routes API (https://developer.tomtom.com/, accessed on 14 May 2021) to assess routes and know their traffic conditions. The requests happened at the same hour when the trips occurred to have similar traffic conditions to those when the flow happened.

5.1.3. Bicycle Data

This data are from the Citi Bike service in New York (https://www.citibikenyc.com/, accessed on 14 May 2021), 2017. This service is the most important bicycle sharing system in the city, and we used 257,498 trips from this dataset. We also used docker's location and availability to calculate route alternatives using shared bicycles. Finally, we used historical trip data to evaluate the bikeability.

5.1.4. Routing Data

We used Google Directions API (https://developers.google.com/, accessed on 14 May 2021) and Uber API (https://developer.uber.com/, accessed on 14 May 2021) to guess public transportation and private hired vehicles routes. The Google Directions API gives the available route segments in public transportation and information regarding bus schedules and live positions (when available) to evaluate waiting times. Uber API is used to obtain fare estimations and waiting times for trips.

5.2. Pre-Processing

Before starting the evaluation, the first step of the analysis consists of splitting the dataset into classes for evaluation. Observing the dataset of Yellow Taxis, we found that an hourly division would be enough to capture its seasonality (see Figure 4). It is possible to observe the variation in the number of trips and passengers according to the hour of the day. Additionally, it is possible to visualize the difference in the curve behavior for weekdays and weekends. Figure 4 shows some relevant areas in the curve, high and low peaks of trips on weekdays, and a peak at the weekend. We made an hourly division to create classes, but the route evaluation used real-time data. These relevant areas were used to evaluate the results (Sections 5.6 and 5.7).

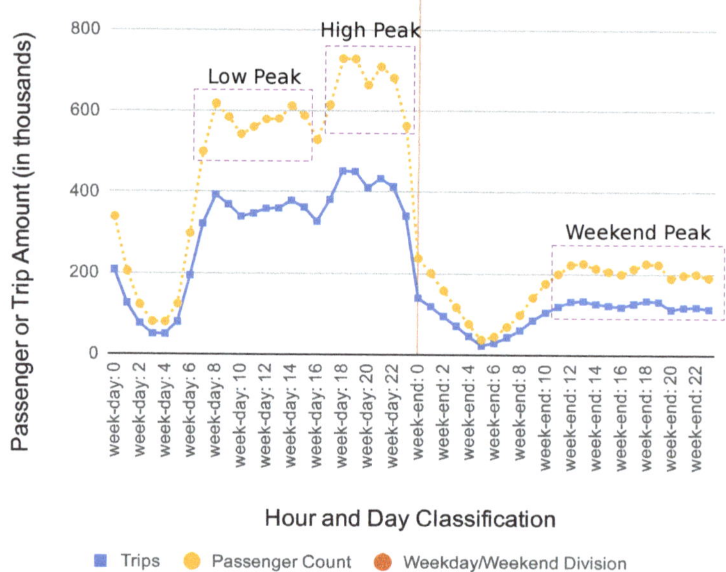

Figure 4. Distribution of trips and passengers in weekdays and weekends per hour; 48 classes were created for every hour of the day in weekdays and weekends [7].

5.3. Data Reduction

We designed a flow clustering technique described in Section 3.3 to reduce the data to be evaluated and keep the relevance of the analysis. Its final step consists of classifying flows between the functional zones. We use a derivative of the curvature equation for this classification. Figure 5 shows the curve of the magnitude of the flows. We smooth the Original Data line, producing the Smoothed line. Later, we use Equation (1) to produce the Curvature line in Figure 5. Its derivative was used to evaluate the maximum, i.e., the Selected Point for the classification (vertical dashed line). Flows with a magnitude above it are considered trending, while the remaining are secondary.

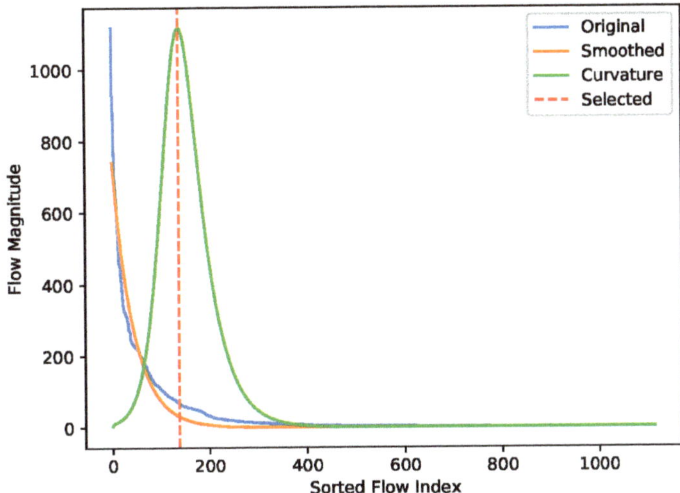

Figure 5. Selection of classification threshold using the derivative of the curvature equation.

Figure 6 shows the classified clustered flows for datasets with and without geo-coordinates. Figure 6a [7] uses the first variation of the flow clustering technique, leading to clusters with a deformed-looking shape. This shape is due to their formation made out of coordinates that were collected from the performed trips. These clusters were obtained from the Yellow Taxis dataset 2016 when geo-coordinates for trip pickup and drop-off points were still available. The clusters shown in Figure 6b,c were tailored—in the anonymization process before data releasing—according to the shapes of some districts and neighborhoods of the city, thus leading to more regular shapes.

Figure 6b,c show a difference in the locations of the departures and arrivals clusters. Yellow Taxis have a historical trend to attend users in Manhattan, while Green Taxis were created to serve the more peripheral zones, being prohibited from starting trips in Manhattan. The focus on covering Manhattan is not strongly present in this sample of flows shown in Figure 6a. Green Taxis are also prohibited from taking passengers to airports, while Yellow Taxis are not, which results in a flow from the bottom-right corner of Figure 6a (where JFK Airport is located) to a cluster near the Brooklyn neighborhood. This is another trend of the Yellow Taxis.

The geo-coordinates allow us to obtain more relevant flows. In contrast, the flows that were obtained with the anonymized dataset resulted in many flows starting and finishing in the same regions. The flows that were obtained in this data reduction phase were used to evaluate the impacts of the hybrid routing, which are described in Section 5.6.

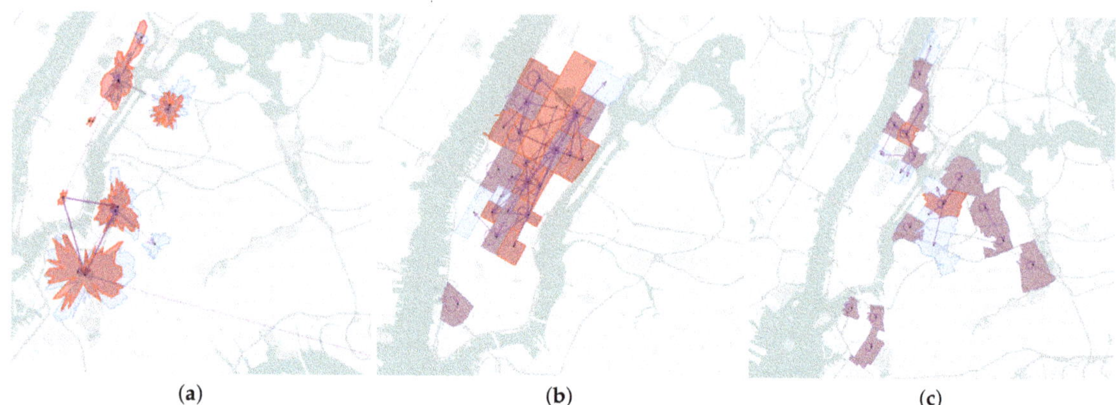
(a) (b) (c)

Figure 6. Clustered flows from Yellow and Green Taxis datasets. (**a**) Yellow Taxis with geo-coordinates. (**b**) Yellow Taxis without geo-coordinates. (**c**) Green Taxis without geo-coordinates.

5.4. Clustering Evaluation

In this section, we compare our proposed clustering methodology with the SSTFC [14]. The objective of applying the clustering in the present study is to allow the identification of the most relevant mobility flows, i.e., groups of trips with similar origin-destination that happen with a high frequency. Figure 7 shows the distribution of clusters given the amount of trips that they contain (their size) excluding trips considered as noise. The clusters were divided into classes with sizes greater than 1000, between 100 and 1000, between 10 and 100, and smaller than 10. We show results for two configurations of each approach, using $\alpha \in \{0.5, 0.05\}$ for SSTFC, and using Polynomial and Exponential model to classify the most relevant flows in our proposal, OD-MFC. The evaluation was performed with all of the clusters from the 48 $\langle weekday,\ hour \rangle$ classes in the non-anonymized NYC Green Taxis dataset.

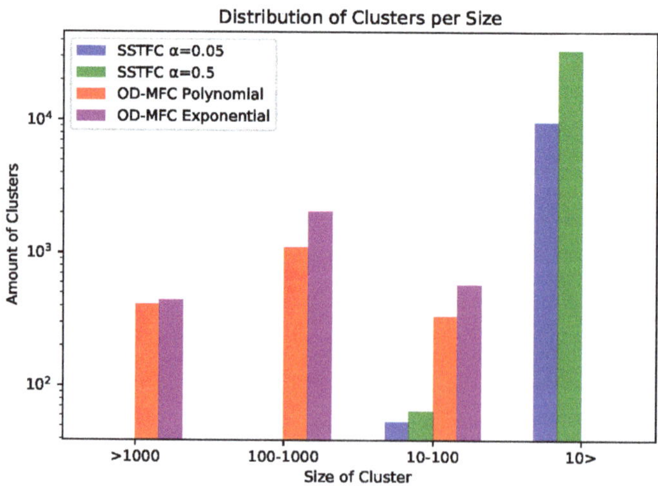

Figure 7. Distribution of flow clusters according to their sizes.

The SSTFC algorithm showed a trend towards identifying a big amount of small clusters, with less than 1% of clusters sized greater than 10 for both parameters studied.

For both scenarios, SSTFC classified more than 99% of the data as noise. OD-MFC could identify more relevant clusters, with all of the selected clusters containing more than 10 trips, and around 60% of the clusters sized between 100 and 1000. OD-MFC classified 71% and 68% of the data as noise, respectively, for Polynomial and Exponential models. In total, we observed 9.7 and 33.4 thousand clusters for SSTFC with α equals to 0.05 and 0.5; and, 1.8 and 3.0 thousand clusters for OD-MFC using Polynomial and Exponential models. Given our objective of selecting relevant flows for evaluation, OD-MFC performed better. Not only more relevant clusters were identified, but also a smaller part of the data was filtered out as noise. There is the possibility that better parameter selection could lead SSTFC towards better results, yet increasing the value of α from 0.05 to 0.5 only increased the amount of smaller clusters identified. Furthermore, initial experiments with smaller α (e.g., 0.005 and 0.0005) did not result in cluster formation, reducing α increases the amount of data classified as noise, as expected. After the clustering step, the average flow of each cluster identified, using the exponential model, was evaluated according to our framework, as described in the following sections.

5.5. Active Transportation Acceptability

The bikeability was obtained using Citi Bike NYC data, as discussed in Section 4.3. We collected data from trips during 2017 and created a cumulative normalized distribution function of the trip lengths (see Figure 8). The curve depicts the probability of some users to use a bicycle given the length of a trip. As its length rises, the number of users drops. This curvature is used as the depreciation factor, in terms of length of trips, when assessing the satisfaction of a given trip segment using bicycles. It is reasonable to use this data since we are using this service as a bicycle provider when evaluating multimodal routes using bike-sharing, i.e., we consider their bike dockers as pickup/drop-off locations when routing trips. Additionally, Figure 8 shows the walkability curve that was used in our work.

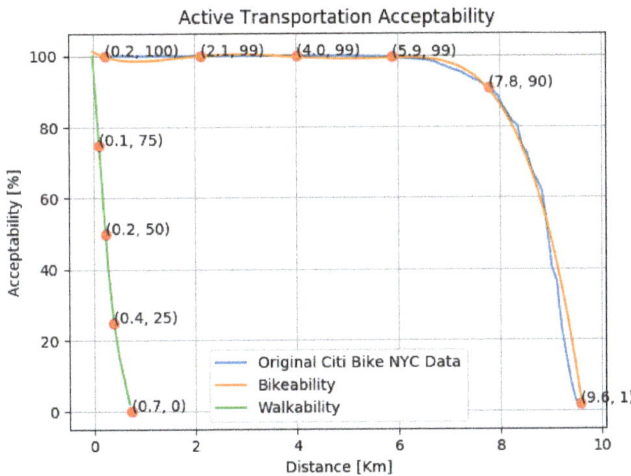

Figure 8. Normalized cumulative distribution function for bike trips using the Citi Bike NYC service in 2017.

To obtain the Original Data curve in Figure 8, trips starting and finishing at the same docker station are dropped, since the bicycles are not being used for mobility. To interpolate the curve and produce Equation (15), we used the highlighted data points (see the bikeability curve in Figure 8). The curve was interpolated as a five-degree polynomial function, whose outcome is a depreciation factor (interval $[0, 1]$) that multiplies the perceived time when using this travel mode. In the figure, the values start to drop fairly around 6 km; before, the impact of the bikeability in the trip's enjoyment is minimal. As shown, the walkability (i.e., Equation (14)) drops quicker than bikeability, as expected due to the use of bicycles to travel longer distances.

5.6. Hybrid Multimodal Routing

The models described in Section 4 were used to compare the route options for the trip flows that were identified according to the data reduction process. Filtering the trip flows to contain only the analyzed traffic scenarios (i.e., Low Peak, High Peak, and Weekend Peak), a total of 1654 trip flows were evaluated. Table 2 shows an overview of the characteristics of these scenarios, with the averages of trips/flow, duration, and length. The results were obtained for the three main outcomes of the model described in Section 4: (i) Generalized Time, (ii) Generalized Cost, and (iii) Opportunity Cost of Generalized Cost. Figure 9a,b show the modes Uber and Transit mean trips that were performed using those modes, while the Hybrid trips consider the mixture of these modes and the use of bicycles.

Table 2. Average flow characteristics for each scenario.

	Trips/Flow	Duration	Length
High Peak Weekdays (18:00–24:00)	5780	13.6 min	1.87 km
Low Peak Weekdays (08:00–17:00)	6453	13.7 min	2.03 km
Weekend Peak Weekends (11:00–02:00)	2378	15.6 min	2.32 km

Figure 9a shows the distribution of the Generalized Time for the main routing options in the three evaluated traffic scenarios. The Generalized Time is a User Experience metric, which outputs an amount of time that the users would perceive passed on average. In order to compute this metric, we use level-of-service and acceptability multipliers to the actual time elapsed. This way, we aim to convert all times elapsed in different modes to a single base in-vehicle time (IVT). Figure 9a shows that the QoE for hybrid routes obtained the best (i.e., smaller) values of medians. This means that, for most trips, the best option is the Hybrid; also, we observe that the distribution of the values is more concentrated for the Hybrid options, meaning more stability on the results. We also note that the variation of the obtained values increased according to the scenario; this is probably due to the resulting average trip length being greater for Low Peak and Weekend Peak than the High Peak.

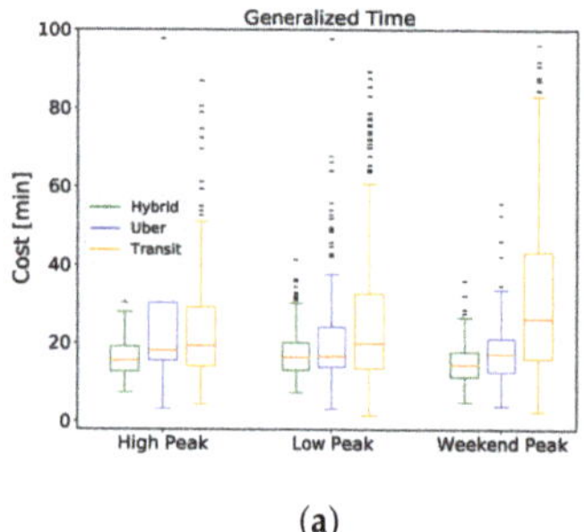

(a)

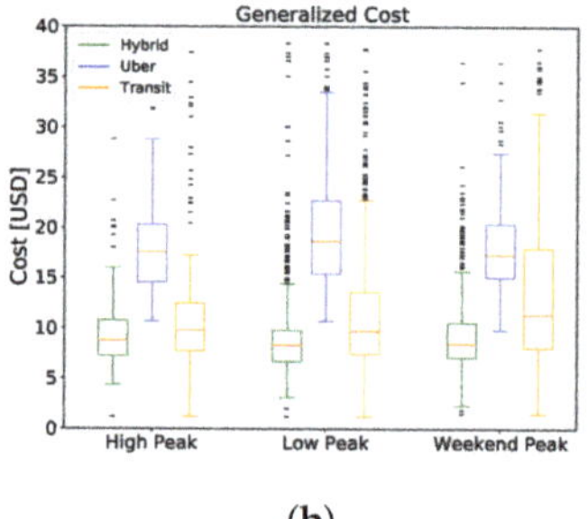

(b)

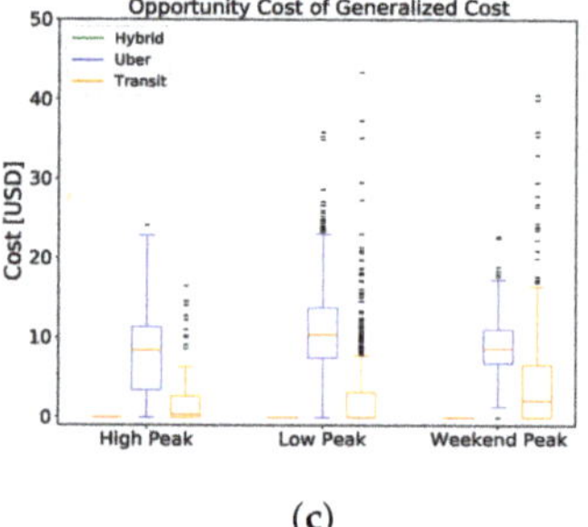

(c)

Figure 9. Relative impact of different users' profiles in the route evaluation. (**a**) Distribution of the Generalized Time of trips. (**b**) Distribution of Generalized Cost of trips. (**c**) Distribution of the Opportunity Cost.

The Generalized Time is used to evaluate the Generalized Cost of a trip (see Figure 9b). The generalized time is used with the average income per hour of full-time year-round Americans [45]. This generalized time is added to the price of the trip to obtain its Generalized Cost (includes price, duration, and user experience). The Hybrid option is the best in terms of Generalized Cost. Because Uber options are more expensive, their values are higher than the Public Transportation ones. We observe the values of Hybrid to be under

Public Transportation. This behavior is probably due to the number of bicycle trips that are cheaper than Public Transportation in New York. When considering the Generalized Cost medians, the average saving values are around USD 10.00, when comparing Hybrid with Uber, and USD 2.50 as compared to Public Transportation in all traffic scenarios.

Figure 9c shows the distribution of Opportunity Cost of choosing an option of a trip w.r.t. the Generalized Cost. This data measures the amount of loss for the decision-maker due to selecting a trip among the available ones. A traditional option is an instance of Hybrid routing. Thus, invariably is the best option in this set, leading the Opportunity Cost to always be zero. The traditional options are not necessarily the best routes, which lead to the distribution of Uber and Public Public Transportation in Figure 9c. Additionally, we consider that bicycle options are in the Hybrid group, causing a rise in the values for the common modes. Regarding the transportation mode changes, the best options of the proposed approach have two changes at maximum—with averages number of changes in each $\langle weekday, hour \rangle$ class under 0.2.

5.7. Route Selection Based on Profiles

In this work, we propose using the Profile Operator (see Section 4.5) to apply our model in individual route recommendations. This operator combines the attention that us given by a user to the experience or price-related metrics. Figure 10 shows the influence of the profiles for the three already described traffic scenarios. In a real-world system, the profile operator should be calibrated with usage data by applying different learning techniques, such as the one that was proposed in the FAVOUR algorithm [5]. Great variations to the duration metric can be observed in Figure 10, this happens since trips with different lengths, which take varied time to be performed, are evaluated. Conversely, the price metric does not vary much, because, for most of the transport modes, the duration does not have a direct impact on the price (e.g., bus, walk, and the initial hiring fee for Ubers).

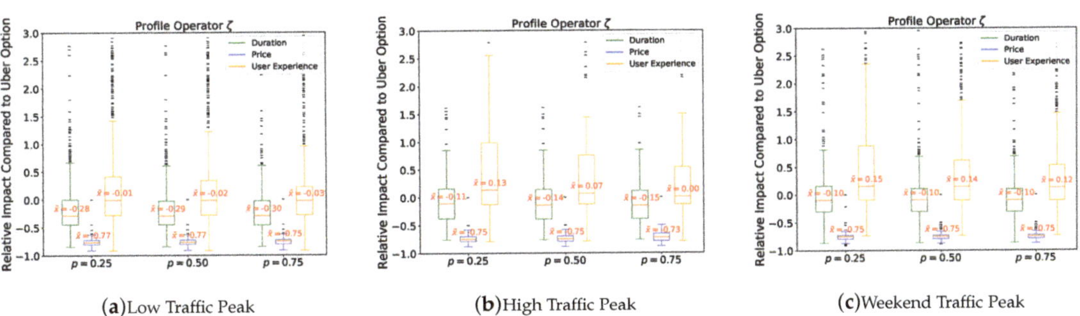

(a) Low Traffic Peak **(b)**High Traffic Peak **(c)**Weekend Traffic Peak

Figure 10. Relative impact of different users' profiles in the route selection.

The profile operator fuses the utility functions after the normalization using a z-score. Next, the ζ profile operator is evaluated, i.e., the route option with the smallest value for each trip is selected for the evaluation—omitting the Uber option, which is used as the baseline. We computed the difference for every metric and then normalized it (divided by the value of the Uber option). We use Equation (19) to calculate the impact in trips, where I is the impact, r' is the Uber option for the trip, and $f(r)$ is the cost function studied (i.e., c_r, t_r, or $T(r)$).

$$I = \frac{\min_{\forall r \in R - \{r'\}} f(r)}{t_{r'}}. \tag{19}$$

For all of the traffic scenarios, we present the expected results. When p increases, the duration and user experience drop, and the cost values rise. This reflects the idea that users expecting a better QoE might pay more for the trip and vice versa. Yet, there are specific situations where the price and duration of trips are reduced together, which results in a cheaper trip with better QoE. For instance, when there is congestion in the streets, leaving a vehicle and choosing a different mode, like a bicycle, may result in a faster trip; since the usage of a hired vehicle is reduced, and the price of the trip decreases.

The median values were plotted over the graphs to facilitate the visualization of this behavior. Most of the observations for all scenarios lie between -1 and 1 (81.7% of the evaluated flows), which means that there are no high gains or losses. All of the median is under the zero mark for the low peak scenario, meaning that gains were obtained in more than 50% of the trips for all metrics. In the High and Weekend peak scenarios, only the metrics duration and price had the same performance, while the median for user experience was a little above the zero mark, even though these median values were not too high.

6. Final Remarks

In this work, we presented a novel flow clustering technique for geo-located and anonymized data and a user experience model for urban trips. These are the main contributions of this paper. We detailed the model and the experiments, so it is possible to replicate this work with other datasets (using the code on Github). Overall, this study showed that hybrid routing is a valuable option for traditional urban routing systems.

The proposed model has some parameters that might be, at times, difficult to calibrate. Some of them can be obtained from datasets that one might have access to, such as the average income of users of public transportation and the bikeability and walkability functions; on the other hand, others cannot be easily obtained, such as the weight for the profile operator. A possible solution, as well as a possible future direction of this study, is to use learning techniques to adapt the weights in the profile operator [5]. Some of the issues were identified when running the experiments, such as the anonymized datasets favored the formation of small-length trips. Thus, some modes have more advantages than others in the analysis. However, we presented an approach to cluster data without anonymization. Besides, the Hybrid option showed to be a viable choice for short trips, and previous results [7] showed that it presents good efficiency in more extended trips. Our solution does not explore the full road network when evaluating routes, but a reduced network suggested by services, such as TomTom Routing and Google Directions. We need to check whether the usage of the full graph (or semi-full depending on the heuristic) would bring gains for our work. This could be achieved by changing the MLGLS layer to another, which implements algorithms, such as the shortest path.

One interesting possible continuation to the present study would be to combine the route recommendation with a central ITS of a smart city. For instance, the mode transition points in our proposal depend on the location of the bus terminals and bike dockers. Once integrated with the central ITS, our approach could be used as a evaluation function to optimize the placement of these stations. To put these ideas in practice, our study would have to be integrated with optimization algorithms with the same purpose [46]. Furthermore, in a integrated ITS scenario, our approach could make use of different strategies to not only keep away from traffic jams, but also avoid their creation. For this, different traffic control mechanisms should be integrated with our method, such as traffic lights and street crossings orchestration control [47].

Author Contributions: Conceptualization, D.O.R., G.M., T.B., A.A.F.L., M.L.M.P., and L.A.V.; methodology, D.O.R., G.M., T.B., A.A.F.L., M.L.M.P., and L.A.V.; software, D.O.R.; validation, D.O.R., G.M., and L.A.V.; formal analysis, D.O.R.; data curation, D.O.R.; writing—original draft preparation, D.O.R., G.M., T.B., A.A.F.L., M.L.M.P., and L.A.V.; writing—review and editing, D.O.R., G.M., T.B., A.A.F.L., M.L.M.P., and L.A.V.; visualization, D.O.R.; supervision, G.M., T.B., A.A.F.L., and L.A.V.; funding acquisition, D.O.R., T.B., L.A.V. All authors have read and agreed to the published version of the manuscript.

Funding: The authors would like to thank the São Paulo Research Foundation (FAPESP), grants #2020/11259-9, #2018/12447-3, #2018/23064-8, #2018/19639-5, #2018/23126-3, and #2015/24494-8.

Institutional Review Board Statement: Not applicable.

Informed Consent Statement: Not applicable.

Data Availability Statement: Publicly available datasets were analyzed in this study. This data can be found at: https://opendata.cityofnewyork.us/, https://developer.tomtom.com/, https://www.citibikenyc.com/, https://www1.nyc.gov/, https://developers.google.com/, and https://developer.uber.com/ all accessed on 14 May 2021.

Conflicts of Interest: The authors declare no conflict of interest.

References

1. Soriano, F.R.; Samper-Zapater, J.J.; Martinez-Dura, J.J.; Cirilo-Gimeno, R.V.; Martinez Plume, J. Smart Mobility Trends: Open Data and Other Tools. *IEEE Intell. Transp. Syst. Mag.* **2018**, *10*, 6–16. [CrossRef]
2. Zhang, D.; Huang, J.; Li, Y.; Zhang, F.; Xu, C.; He, T. Exploring Human Mobility with Multi-Source Data at Extremely Large Metropolitan Scales. In Proceedings of the 20th Annual International Conference on Mobile Computing and Networking, Maui, HI, USA, 7–11 September 2014; pp. 201–212. [CrossRef]
3. Sweeney, S.; Ordóñez-Hurtado, R.; Pilla, F.; Russo, G.; Timoney, D.; Shorten, R. A Context-Aware E-Bike System to Reduce Pollution Inhalation While Cycling. *IEEE Trans. Intell. Transp. Syst.* **2019**, *20*, 704–715. [CrossRef]
4. Quercia, D.; Schifanella, R.; Aiello, L.M. The Shortest Path to Happiness: Recommending Beautiful, Quiet, and Happy Routes in the City. In Proceedings of the 25th ACM Conference on Hypertext and Social Media, Santiago, Chile, 1–4 September 2014; pp. 116–125. [CrossRef]
5. Campigotto, P.; Rudloff, C.; Leodolter, M.; Bauer, D. Personalized and Situation-Aware Multimodal Route Recommendations: The FAVOUR Algorithm. *IEEE Trans. Intell. Transp. Syst.* **2017**, *18*, 92–102. [CrossRef]
6. Houshmand, A.; Cassandras, C.G. Eco-Routing of Plug-In Hybrid Electric Vehicles in Transportation Networks. In Proceedings of the 2018 21st International Conference on Intelligent Transportation Systems (ITSC), Maui, HI, USA, 4–7 November 2018; pp. 1508–1513. [CrossRef]
7. Rodrigues, D.O.; Fernandes, J.T.; Curado, M.; Villas, L.A. Hybrid Context-Aware Multimodal Routing. In Proceedings of the 2018 21st International Conference on Intelligent Transportation Systems (ITSC), Maui, HI, USA, 4–7 November 2018; pp. 2250–2255. [CrossRef]
8. Train, K.E. *Discrete Choice Methods with Simulation*, 2nd ed.; Cambridge University Press: Cambridge, UK, 2009.
9. Dotoli, M.; Zgaya, H.; Russo, C.; Hammadi, S. A Multi-Agent Advanced Traveler Information System for Optimal Trip Planning in a Co-Modal Framework. *IEEE Trans. Intell. Transp. Syst.* **2017**, *18*, 2397–2412. [CrossRef]
10. Hong, Z.; Chen, Y.; Mahmassani, H.S. Recognizing Network Trip Patterns Using a Spatio-Temporal Vehicle Trajectory Clustering Algorithm. *IEEE Trans. Intell. Transp. Syst.* **2018**, *19*, 2548–2557. [CrossRef]
11. Pang, J.; Huang, J.; Yang, X.; Wang, Z.; Yu, H.; Huang, Q.; Yin, B. Discovering Fine-Grained Spatial Pattern From Taxi Trips: Where Point Process Meets Matrix Decomposition and Factorization. *IEEE Trans. Intell. Transp. Syst.* **2018**, *19*, 3208–3219. [CrossRef]
12. Zhong, G.; Wan, X.; Zhang, J.; Yin, T.; Ran, B. Characterizing Passenger Flow for a Transportation Hub Based on Mobile Phone Data. *IEEE Trans. Intell. Transp. Syst.* **2017**, *18*, 1507–1518. [CrossRef]
13. El Mahrsi, M.K.; Côme, E.; Oukhellou, L.; Verleysen, M. Clustering Smart Card Data for Urban Mobility Analysis. *IEEE Trans. Intell. Transp. Syst.* **2017**, *18*, 712–728. [CrossRef]
14. Yao, X.; Zhu, D.; Gao, Y.; Wu, L.; Zhang, P.; Liu, Y. A Stepwise Spatio-temporal Flow Clustering Method for Discovering Mobility Trends. *IEEE Access* **2018**, *6*, 44666–44675. [CrossRef]
15. Birant, D.; Kut, A. ST-DBSCAN: An Algorithm for Clustering Spatial–temporal Data. *Data Knowl. Eng.* **2007**, *60*, 208–221. [CrossRef]
16. Ester, M.; Kriegel, H.P.; Sander, J.; Xu, X. A Density-based Algorithm for Discovering Clusters in Large Spatial Databases with Noise. In Proceedings of the Second International Conference on Knowledge Discovery and Data Mining (KDD-96), Portland, OR, USA, 2–4 August 1996; pp. 226–231.
17. Amar, H.M.; Drawil, N.M.; Basir, O.A. Traveler Centric Trip Planning: A Behavioral-Driven System. *IEEE Trans. Intell. Transp. Syst.* **2016**, *17*, 1521–1537. [CrossRef]
18. Ben-Akiva, M.; Bierlaire, M. Discrete Choice Models with Applications to Departure Time and Route Choice. In *Handbook of Transportation Science*; Hall, R.W., Ed.; Springer: New York, NY, USA, 2003; pp. 7–37. [CrossRef]
19. Hrnčíř, J.; Žilecký, P.; Song, Q.; Jakob, M. Practical Multicriteria Urban Bicycle Routing. *IEEE Trans. Intell. Transp. Syst.* **2017**, *18*, 493–504. [CrossRef]
20. Campello, R.J.G.B.; Moulavi, D.; Zimek, A.; Sander, J. Hierarchical Density Estimates for Data Clustering, Visualization, and Outlier Detection. *ACM Trans. Knowl. Discov. Data* **2015**, *10*, 1–51. [CrossRef]
21. Thorpe, J.A. Curvature of Plane Curves. In *Elementary Topics in Differential Geometry*; Springer: New York, NY, USA, 1979; pp. 62–67. [CrossRef]
22. McInnes, L.; Healy, J. Accelerated Hierarchical Density Based Clustering. In Proceedings of the IEEE International Conference on Data Mining Workshops, New Orleans, LA, USA, 18–21 November 2017; pp. 33–42. [CrossRef]

23. Ueda, K.; Yamashita, N. On a Global Complexity Bound of the Levenberg-Marquardt Method. *J. Optim. Theory Appl.* **2010**, *147*, 443–453. [CrossRef]
24. Lorimer, T.; Held, J.; Stoop, R. Clustering: How Much Bias Do We Need? *Philos. Trans. R. Soc. A* **2017**, *375*, 1–18. [CrossRef] [PubMed]
25. Latapy, M.; Fiore, M.; Ziviani, A. Link Streams: Methods and Applications. *Comput. Netw.* **2019**, *150*, 263–265. [CrossRef]
26. Bentley, J.L. Multidimensional Binary Search Trees Used for Associative Searching. *ACM Commun.* **1975**, *18*, 509–517. [CrossRef]
27. Kivelä, M.; Arenas, A.; Barthelemy, M.; Gleeson, J.P.; Moreno, Y.; Porter, M.A. Multilayer Networks. *J. Complex Netw.* **2014**, *2*, 203–271. [CrossRef]
28. Sharifzadeh, M.; Shahabi, C. VoR-Tree: R-trees with Voronoi Diagrams for Efficient Processing of Spatial Nearest Neighbor Queries. *Proc. VLDB Endow.* **2010**, *3*, 1231–1242. [CrossRef]
29. Zheng, Y.U. Trajectory Data Mining: An Overview. *ACM Trans. Intell. Syst. Technol.* **2015**, *6*, 1–41. [CrossRef]
30. Rodrigues, D.O.; Boukerche, A.; Silva, T.H.; Loureiro, A.A.F.; Villas, L.A. SMAFramework: Urban Data Integration Framework for Mobility Analysis in Smart Cities. In Proceedings of the 20th ACM Int'l Conference on Modelling, Analysis and Simulation of Wireless and Mobile Systems, Miami, FL, USA, 21–25 November 2017; pp. 227–236. [CrossRef]
31. Buchanan, J.M. Opportunity Cost. In *The World of Economics*; Eatwell, J., Milgate, M., Newman, P., Eds.; Palgrave Macmillan: London, UK, 1991; pp. 520–525. [CrossRef]
32. Bruzelius, N.A. Microeconomic Theory and Generalised Cost. *Transportation* **1981**, *10*, 233–245. [CrossRef]
33. Wardman, M.; Toner, J. Is Generalised Cost Justified in Travel Demand Analysis? *Transportation* **2020**, *47*, 75–108. [CrossRef]
34. Cherlow, J.R. Measuring Values of Travel Time Savings. *J. Consum. Res.* **1981**, *7*, 360–371. [CrossRef]
35. Abrantes, P.A.; Wardman, M.R. Meta-analysis of UK Values of Travel Time: An Update. *Transp. Res. Part A Policy Pract.* **2011**, *45*, 1–17. [CrossRef]
36. Ye, R.; Titheridge, H. Satisfaction with the Commute: The Role of Travel Mode Choice, Built Environment and Attitudes. *Transp. Res. Part D* **2017**, *52*, 535–547. [CrossRef]
37. Moura, F.; Cambra, P.; Gonçalves, A.B. Measuring Walkability for Distinct Pedestrian Groups with a Participatory Assessment Method. *Landsc. Urban Plan.* **2017**, *157*, 282–296. [CrossRef]
38. Nielsen, T.A.S.; Skov-Petersen, H. Bikeability—Urban Structures Supporting Cycling. Effects of Local, Urban and Regional Scale Urban form Factors on Cycling from Home and Workplace Locations in Denmark. *J. Transp. Geogr.* **2018**, *69*, 36–44. [CrossRef]
39. Schaap, N.; Harms, L.; Kansen, M.; Wüst, H. *Cycling and Walking: The Grease in Our Mobility Chain*; KiM Netherlands Institute for Transport Policy Analysis: Den Haag, The Netherlands, 2016.
40. Yang, Y.; Diez-Roux, A.V. Walking Distance by Trip Purpose and Population Subgroups. *Am. J. Prev. Med.* **2012**, *43*, 11–19. [CrossRef]
41. Ryus, P.; Danaher, A.; Walker, M.; Nichols, F.; Carter, B.; Ellis, E.; Cherrington, L.; Bruzzone, A. Quality of Service, Ridership and Service Cost. In *Transit Capacity and Quality of Service Manual*; Transit Cooperative Research Program: Washington, DC, USA, 2013.
42. De Souza, A.M.; Braun, T.; Villas, L. Efficient Context-Aware Vehicular Traffic Re-Routing Based on Pareto-Optimality: A Safe-Fast Use Case. In Proceedings of the 2018 21st International Conference on Intelligent Transportation Systems (ITSC), Maui, HI, USA, 4–7 November 2018; pp. 2905–2910. [CrossRef]
43. Kraus, S.; Arkin, R.C. *Strategic Negotiation in Multiagent Environments*; MIT Press: Cambridge, MA, USA, 2001.
44. Briggs, R. Normative Theories of Rational Choice: Expected Utility. In *The Stanford Encyclopedia of Philosophy*; Zalta, E., Ed.; Center for the Study of Language and Information (CSLI): Stanford, CA, USA, 2017.
45. Survey, A.C. Selected Economic Characteristics: 2013–2017 ACS 5-Year Estimates, 2018. Available online: https://www.census.gov/programs-surveys/acs/technical-documentation/table-and-geography-changes/2017/5-year.html (accessed on 12 May 2021).
46. Arabi, M.; Beheshtitabar, E.; Ghadirifaraz, B.; Forjanizadeh, B. Optimum Locations for Intercity Bus Terminals with the AHP Approach—Case Study of the City of Esfahan. *Int. J. Environ. Ecol. Eng.* **2015**, *9*, 545–551.
47. Munoz-Organero, M.; Ruiz-Blaquez, R.; Sánchez-Fernández, L. Automatic detection of traffic lights, street crossings and urban roundabouts combining outlier detection and deep learning classification techniques based on GPS traces while driving. *Comput. Environ. Urban Syst.* **2018**, *68*, 1–8. [CrossRef]

Article

Methodology of Functional and Technical Evaluation of Cooperative Intelligent Transport Systems and Its Practical Application

Zdeněk Lokaj, Martin Šrotýř, Miroslav Vaniš * and Michal Mlada

Faculty of Transportation Sciences, Czech Technical University in Prague, 110 00 Prague, Czech Republic; lokaj@fd.cvut.cz (Z.L.); srotyr@fd.cvut.cz (M.Š.); mladamic@fd.cvut.cz (M.M.)
* Correspondence: vanismir@fd.cvut.cz

Abstract: In the area of smart cities, great emphasis is placed on many different fields such as energetics, information systems, and transportation. All of these should lead to a simplification of life thanks to smart technologies. If we talk about the transportation field, the main issues related to this area are safety, traffic efficiency, or the environment. Another condition is the successful acceptance of any new technology by its users. Cooperative systems prove to be a suitable solution for these issues, especially in urban areas. Today, pilot implementations of cooperative systems in European countries are being carried out. However, before they are put into full operation, they need to be tested, evaluated, and assessed. This article focuses on the latter two points, i.e., evaluation and assessment of the cooperative systems. For this purpose, a methodology was created, which describes the procedure chosen in the evaluation and assessment of cooperative systems in the Czech Republic and a demonstration of its use by example. The methodology is focused on three main areas, which in this case are functional evaluation, user acceptance, and impact assessment. For the area of user acceptance, the main source was questionnaires, impact assessment relied on measured data while functional evaluation was based on discussions with the drivers, evaluating the cooperative systems, the measured data, and the expert observations. All collected and measured data were then processed and some of the results of the evaluation of the selected service are presented at the end of this article.

Keywords: cooperative systems; evaluation methodology; telematics; functional evaluation; impact assessment; user acceptance

Citation: Lokaj, Z.; Šrotýř, M.; Vaniš, M.; Mlada, M. Methodology of Functional and Technical Evaluation of Cooperative Intelligent Transport Systems and Its Practical Application. *Appl. Sci.* **2021**, *11*, 9700. https://doi.org/10.3390/app11209700

Academic Editor: Leon Rothkrantz

Received: 9 September 2021
Accepted: 13 October 2021
Published: 18 October 2021

Publisher's Note: MDPI stays neutral with regard to jurisdictional claims in published maps and institutional affiliations.

1. Introduction

Today, Cooperative Intelligent Transport Systems (C-ITS) are becoming one of the main new technologies in transport. Their main goal is to increase traffic safety, traffic flow, and to reduce the negative impact on the environment through the acquisition of real-time targeted traffic information and rapid response to it.

C-ITS are based on the exchange of data between two vehicles via OBU (On-board unit) and between vehicles (OBU) and other elements such as RSU (Road-side unit), BO (Back-office), or mobile phones.

The basic principle of C-ITS is to send up-to-date information of various kinds such as a warning or a temporary event via C-ITS messages sufficiently in advance. These warnings can be divided according to the type of information they provide. A typical example might be a road works warning, a traffic jam ahead warning, or a weather condition warning. These events are generally called use cases. More information about C-ITS and its use cases can be found in [1–3].

In terms of this article, it is important that certain use-cases are defined in C-ITS, about which the C-ITS user is informed in advance where he can then decide how to react to this event. Before C-ITS is put into full operation, C-ITS must not only be tested but also evaluated and assessed.

One of the largest pilot projects dealing with individual aspects of C-ITS in Europe is called C-Roads. The C-Roads is a joint initiative of states with road operators and other partners involved in the testing and implementation of C-ITS with a focus on interoperability and harmonization at the European level. C-Roads addresses all fundamental issues related to C-ITS, including evaluation and assessment, which are the main subject of this article.

At the national level, C-ITS can be deployed and tested at selected pilot sites. Furthermore, there can be an evaluation and assessment of these systems in the form of selected use-cases.

The evaluation process could be deployed after the successful field testing is completed. Therefore, the evaluation is no longer focused on standards and specifications within the evaluation. The evaluation process is performed to determine the benefits of C-ITS.

This article deals with the methodology of C-ITS evaluation and assessment in the Czech Republic and demonstrates its use in practice in the form of the performed evaluations.

The presented methodology and example results have the advantage of working with real data from an existing C-ITS system in the field with test drivers, not just simulation or data from test circuits. Furthermore, the concept of two runs (with and without C-ITS technology) proved to be useful to compare these two approaches. The inclusion of additional sensors to evaluate the runs is also beneficial.

State of the Art

The essential document dealing with evaluation procedures is the FESTA Handbook [4]. It is a guideline document with intentions to gather all knowledge from experts, stakeholders, workshops, and seminars about Field operational tests (FOT) to create a common methodology. FOT follows evaluation and assessment methods for driver support systems testing newly developed or implemented systems to provide real-world impact and benefits. The evaluation processes are based on the so-called V diagram, which describes the individual steps recommended within the FOT.

The evaluation framework used in the InterCor project is based on the above-mentioned document [5]. This framework defines "what" needs to be evaluated. It defines the artifacts for input and output for evaluation, such as the research questions and hypotheses to be tested and answered, and the required key performance indicators and measurements from pilot data. In this framework, it is also stated that it is necessary to adjust the recommendations provided by the FESTA handbook according to the evaluated use-cases.

Another project dealing with C-ITS and thus its evaluation is the Nordicway project [6], which includes Finland, Sweden, and Denmark. The evaluation here was addressed in several different areas such as service ecosystem, user acceptance, or quality of service. For the methodology used in Nordicway 2, some steps defined in FESTA were used and adapted for the needs of evaluation.

The next project to be mentioned is called COOPERS [7,8]. Impact assessment is first determined on simulators, where it is possible to test some risks of the situation. The impact assessment was focused on the speed profiles, acceleration, lane-changing behavior, and car-following gaps. The drivers also filled in questionnaires after the drive aiming at the usefulness and, readiness of the information provided by C-ITS.

There are also other projects dealing with cooperative systems, including their evaluation, such as DRIVE C2X [9] and SCOOP [10]. For the SCOOP project, for example, the evaluation focuses not only on vehicle drivers but also on road operators, while the DRIVE C2X project focuses on another area of journey quality evaluation.

Another possibility of evaluation is the use of modeling tools. This approach is used, for example, in Italy [11] in the evaluation of platooning and its effects on safety or the environment.

In addition to the above-mentioned documents and articles, research is also moving towards modification and expansion of the FESTA handbook, e.g., [12,13].

In Australia, they are also working on evaluation based on cooperative awareness messages to assess safety impact [14].

Article [15] deals with a two-stage study including a structured interview with guidelines for assessing acceptance factors and an online study for their evaluation. The study is also focused on facilities that can be provided by the cooperative driver assistant systems in the future like braking, lane change, etc. The interesting conclusion reflects the different opinions of car and truck drivers on C-ITS problematic.

Article [16] examines the user's consent to an automatic overtaking system based on cooperative systems. The user acceptance is focused on usefulness, ease of use, enjoyment, safety, etc.

Article [17] is focused on transport factors that affect cooperative systems. These are solved and measured using simulation tools. Challenges and difficulties are also mentioned here, such as choosing the right indicators or defining proper scenarios for determining these factors.

The next article [18] is aimed at the effect of HMI on driver perception. It compares two basic HMI modes: one-stage, i.e., informing about only one event (the one with the highest priority), or three-stage, i.e., informing about several events. Testing took place on a driving simulator. The bottom line is that drivers would prefer a 3-stage warning system.

Another study [19] deals with the design of an interface to the infotainment systems in a vehicle. It contains an extensive study of articles dealing with this issue, focusing on the positives and negatives of these systems in order to mitigate the negatives.

The last but not least mentioned document is the Evaluation and Assessment plan [20] provided by the European group within the C-Roads focused on the evaluation and assessment created by the representatives from each C-Roads member state. This document provides recommendations and advice to evaluate and assess C-ITS in the correct way and is focused on the user acceptance and the impact assessment in the following areas: safety, traffic efficiency, environment. It is assumed that this plan will be further extended to other areas based on ongoing evaluations.

2. Materials and Methods

2.1. Evaluation Methodology

Each Member State within C-Roads approaches evaluation and assessment differently due to different specificities including the different types of use-cases or conditions that can be created on the pilot site in order to perform the evaluation.

The evaluation methodology described below deals with these specifics and is in accordance with both the FESTA handbook [4] and the evaluation and assessment plan of C-Roads [20].

The methodology first focusing on the whole evaluation process consists of

- Evaluation preparation,
- Pilot site evaluation,
- Evaluation and assessment.

In the Czech Republic, the part of evaluation and assessment covers only the following areas:

- User acceptance,
- Impact assessment,
- Functional evaluation.

All the previously mentioned points are gradually analyzed and finally shown by an example.

2.1.1. Evaluation Preparation

The preparation of the evaluation requires several points:

- cooperation with all relevant partners in the project on:
    - selection of an area for evaluation within the pilot site,
    - selection of specific use-case(s),
    - evaluation date,
    - preparation of evaluation route,
    - specific requirements related to the service concerned,
    - provision of drivers to independently assess C-ITS,
- preparation of specific questionnaires related to the evaluated services,
- provision and preparation of a special "evaluation" vehicle equipped with HMI and other sensors for future assessment.

Drivers addressed to carry out evaluation routes should be diverse in age, sex, mileage covered, driving experience, etc. This will make the evaluation assessment more relevant.

2.1.2. Pilot Site Evaluation

This phase takes place after all points from the evaluation preparation part are agreed upon and deployed. Appropriate events should be arranged at the pilot site or on the evaluation route. There is always an effort to create the event (represented by a use case in terms of C-ITS) in such a way that its location is not affected by other circumstances and that it is really taking a place in a given area. Inappropriate circumstances could significantly affect the results of the evaluation in a negative sense.

The typical course at the pilot site for the evaluation driver is as follows:

- meeting of drivers at the agreed time and place,
- filling in the relevant pre-ride questionnaires,
- first evaluation ride (with C-ITS off),
- second evaluation ride (with C-ITS on),
- filling in the relevant post-ride questionnaires.

More information about the questionnaires can be found in Section 2.1.3. User Acceptance in this paper.

The aim of the evaluation is to find out the benefits resulting from the use of C-ITS services but also their shortcomings, to gain knowledge about the possibility of future improvements. For this reason, ideally, the driver should travel the evaluation route twice. In the first case, he will not have any C-ITS elements, mainly HMI, to inform him about the event. During the second evaluation ride, these systems will be switched on and the driver will be informed of the event in advance via C-ITS. The goal of this procedure is to compare drivers' reactions to the traffic events (use cases) using C-ITS services and the baseline behavior without any service. Impact assessment addresses these issues in detail in Section 2.1.3. Impact Assessment and its example makes up part of Section 3.1.3. Impact Assessment.

Evaluation results can be affected by many factors. Such factors include, for example, weather changes (rain, strong wind, direct sunlight), light visibility (day/night), situation clarity (visibility at the intersection, objects blocking the view), used car, traffic restrictions (speed, priority), traffic flow (low/high), and road topology (urban/extra-urban). These factors can influence driver behavior and lead to misleading results. For this reason, it is recommended that the evaluation be performed under the same conditions on the same test circuit for all drivers. In this way, the influence of random factors can be reduced and thus approached more precisely in determining the true impact of the tested system on the driver.

2.1.3. Evaluation and Assessment

The last phase is the evaluation and assessment of all obtained data in three evaluation areas: user acceptance, safety impact, and functional evaluation.

User Acceptance

Four basic questionnaires were prepared for user acceptance (in accordance with WG3):

- Driver's profile,
- General questions regarding C-ITS,
- Questionnaires related to a specific use-case:
    - Pre-ride,
    - Post-ride.

For most questions in the driver's profile questionnaire, the driver ticked off the appropriate option. The questions are from the following areas:

- Type of driver,
- Sex,
- Age (in ranges),
- Education,
- Number of citizens in hometown (in range),
- Type of a driving license,
- The length of driver's license ownership,
- Number of driven kilometers annually,
- Frequency of driving vehicle,
- The number of (caused and uncaused) traffic accidents, penalty points,
- Current sources of traffic information,
- Preferred traffic information based on C-ITS.

General questions related to C-ITS focused mainly on two areas: willingness to pay for these services (regularly or once) and the opinion of drivers on the HMI and its distraction while driving.

Specific pre-ride questionnaires related to a given use-case do not always have the same questions, however, they focus on driver perceived perception in the following areas:

- Safety,
- Overview of a situation,
- Comfort of driving.

Specific post-ride questionnaires aim at all previous areas and further:

- Usefulness of the service,
- Satisfaction with the information obtained.

Samples of the results are shown in Section 3.1.3 User Acceptance.

Impact Assessment

The impact assessment for different use-cases in C-Roads follows the recommendation and general guidelines in the Evaluation and assessment plan [20], proposed by WG3 within the C-Roads platform. These specifications were designed to unify the process of evaluation and assessment and the results of the individual sub-states at their pilot sites in order to be as transparent and comprehensible as possible for all C-ITS stakeholders. Furthermore, it proposes general recommendations at the level of individual use-cases but leaves the method of implementation to individual states.

In the Evaluation and assessment plan [20], the main three impact areas covering the expected benefits and impact of the C-ITS services are in line with national and European policies to increase safety and reduce the environmental impact of transportation. Each area of impact brings a slightly different view of the evaluation of changes in driver behavior and is linked to subsequent analysis and individual statistical methods that use the captured data.

Within the proposed methodology, impact area safety was selected as the main part. For each use case, an analytical method was chosen based on the true nature of each evaluated use case and the ability of the recording equipment to capture the difference in driver behavior using C-ITS as opposed to driving without C-ITS. The main comparison of individual passage for all use-cases was the comparison of speeds and comparison of accelerations. To determine the significance of the differences between the two mean values (situations with and without C-ITS), a two-sided, paired T-Test was used when comparing speed and acceleration. Such a test evaluates whether there is a possibility that the difference in the mean value of the two selections is due to chance. Levene's test was performed on the same comparison of speed and acceleration to determine if there was a statistical difference in the variances with and without C-ITS.

Functional Evaluation

Functional evaluation in this methodology covers aspects of evaluation that are based on the experience of evaluators, evaluation drivers, and other stakeholders. It also deals with some problematic properties of use-cases. Functional evaluation is therefore divided into three areas:

- lessons learned,
- HMI,
- quality of service.

Lessons learned are primarily designed from the user's point of view. This means that what the driver lacked in the use-case, he would like to improve, etc. It also focuses on improving the use-case in terms of implementation.

In the HMI section, improvements to the C-ITS message presentation are addressed.

Quality of service then focuses on information related to the message parameters, e.g., whether the range was adequate, the relevant zone set correctly, etc.

3. Results

3.1. Example Application of the Methodology

This paper showed a brief theoretical description of the evaluation methodology in the previous section. In this section, we describe the application of the methodology using a specific example, including a presentation of the results from this evaluation. Public transport vehicle crossing (PTVC) was chosen as an exemplary C-ITS use-case for presenting the results of the C-Road CZ evaluation at the pilot sites.

The goal of the PTVC use-case is to notify road users in advance about trams (or other public transport vehicles) crossing the road on the expected route. This application is useful especially in problematic locations, where trams cross the roadway without the aid of traffic lights.

3.1.1. Initial Conditions

For evaluation purposes we possess a test vehicle that is equipped with OBU Commsignia ITS OB4, GPS data logger, OBD2 CAN bus logging device, and HMI. As part of the evaluation process, we also use the C-ITS SIM tool, which is described in more detail in article [21]. This SIM tool was mainly used as the primary source for capturing the communication of C-ITS units to capture data of individual driver's rides. The GPS logger (Canmore GP-102+) tracker and CAN-BUS data logger (CANedge1) were used as additional recording devices. These recording devices were used mainly as a backup in case of failure of the main data source and for other experimental purposes. More information about the C-ITS SIM recording device is described in article [21].

3.1.2. Evaluation Preparation + Pilot Site Evaluation

A timetable was set in advance for each test location and potential respondents were approached to participate in the evaluation. The goal was always to obtain a minimum of 15 diverse respondents for the evaluation of each use-case. Numbers varied for different locations as well as use-cases. The time for evaluation was always determined based on local conditions so that the evaluation would not be affected by local traffic problems or abnormalities.

The presented example of the PTVC use-case was implemented in the cities of Ostrava and Pilsen. Testing had already been successfully performed at these pilot sites.

The first task was to choose the location where the evaluation will take place. The city of Ostrava with a tram crossing in the place where the arriving vehicle does not have the possibility to see the approaching tram from a distance was chosen for the evaluation.

The evaluation route was consulted with the relevant partners (DPO-Public Transport in Ostrava) and is shown in Figure 1. The meeting point for evaluation drivers was Point 1. Here drivers filled out the driver's profile questionnaire and pre-ride questionnaire. Then the driver went on an evaluation drive to point 2, where he also passed point 3 indicating the crossing with a tram. This passing was done without C-ITS. The evaluators at the site (point 3) had an overview of the approaching trams, and thanks to this, the driver was always instructed by the co-driver to start from point 2 in order to meet the tram in the crossing area. The main advantage of this use-case is that when the driver's visibility to the tram is reduced, as in this situation, the driver has information about the arrival of the tram before he sees it. Then his ride continued back to point 1 where he filled out the post-ride questionnaire and his participation in the evaluation ended. This was followed by the whole process again with another driver until further respondents were available or until it was technically and timewise feasible to carry out the evaluation. The length of the route for one participant was approximately 3 km and the journey time was approximately 13 min, including the waiting time for the arrival of the tram.

Figure 1. Evaluation route within the Public transport vehicle crossing use-case.

The evaluation took place at the same tram crossing for all drivers and on two consecutive days, always from 9 am to 1 pm. As the evaluation was held on sunny days in the summer, the light visibility was comparable during all evaluation runs. Speed restrictions and traffic restrictions were the same on both days of testing. The tram crossing is located in the connecting street on the main street, so the traffic was constant and minimal during the whole day. The test vehicle used in the evaluation by all evaluated drivers was Ford C-Max also shown in Figure 2 with the mentioned tram crossing.

Figure 2. The test vehicle on the tram crossing under evaluation ride (point 3 on the map above).

3.1.3. Evaluation Assessment

The assessment results are divided according to the evaluation areas into user acceptance, safety impact, and functional evaluation.

User Acceptance

User acceptance results were provided using different types of graphs. The answers to most questions from the driver's profile questionnaires are shown in pie or bar graphs. The diversity of drivers in terms of age, mileage covered, or the size of the city in which the driver lives can be seen in Figures 3–5.

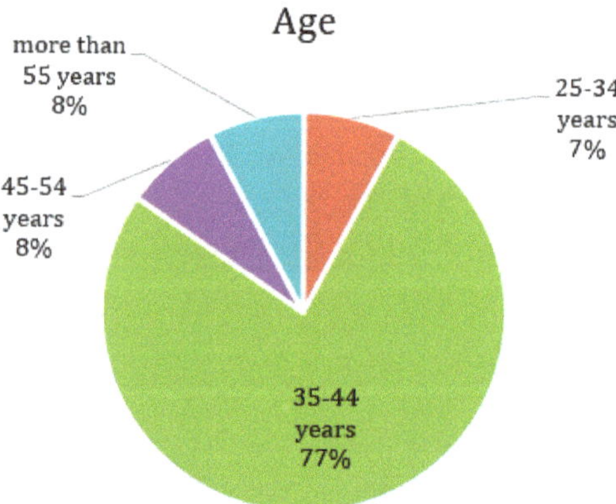

Figure 3. Representation of drivers in terms of age.

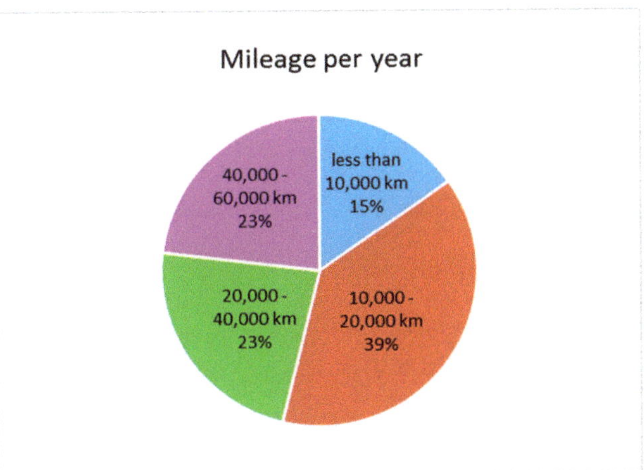

Figure 4. Representation of drivers in terms of mileage per year.

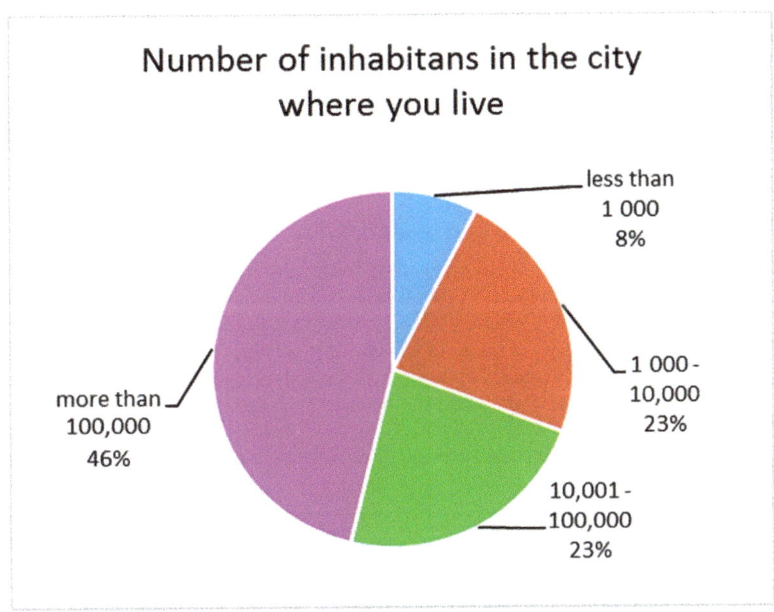

Figure 5. Representation of drivers in terms of the size of the driver's city.

In addition, the driver was also asked as to what traffic information he would like to receive while driving. The results are shown in Figure 6 where the driver wants to be mainly informed about traffic jams, but also about specific situations. This is very positive in terms of the C-ITS. C-ITS is focused on this type of information. On the other hand, drivers are not very interested in lane information.

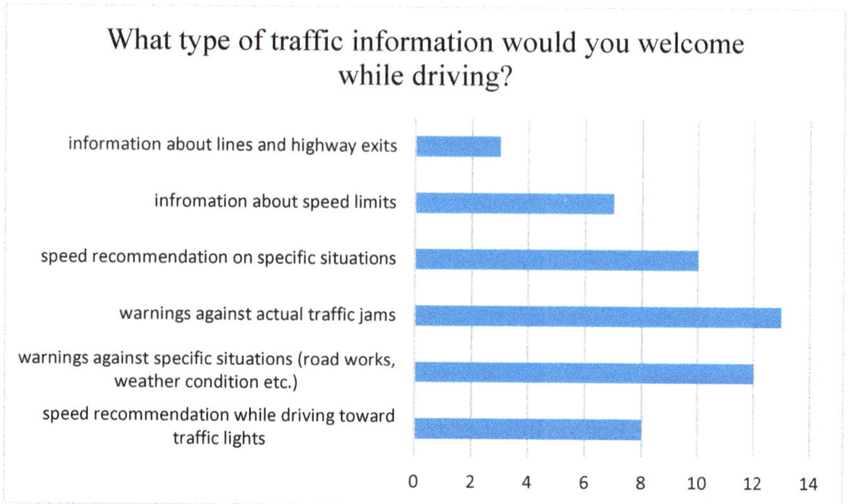

Figure 6. Types of information that drivers would welcome while driving.

In the other types of questionnaires, the answers to the questions asked were in the range:

1. strongly disagree,
2. disagree,
3. neutral attitude,
4. agree,
5. strongly agree.

In Table 1 we see the answers to the questions obtained before the ride. Drivers had an overall neutral attitude towards HMI, however, this value had a relatively large variance. This result is because drivers are diverse and that not every driver could imagine what a ride with an HMI would look like. The drivers also agreed (normally or strongly) to be informed about the crossing in advance when a tram is approaching. They also think that the safety and the overview of the situation near the crossing will increase.

Table 1. The results from pre-ride questionnaires including specific questions on general and the use-case Public Transport Vehicle Crossing.

Pre-Ride Questions Related to	Mean Value	Median Value	Mode	Variance	Standard Deviation
HMI distraction	3.31	3	2	1.23	1.11
Information about trams approaching the crossing	4.69	5	5	0.23	0.48
Safety increase	4.62	5	5	0.26	0.51
Overview of the situation	4.31	4	4	0.40	0.63

The post-ride results regarding the information registration differ, see Figure 7. Here we can also see the different perceptions of drivers in terms of message registration. This again stems from the diversity of drivers. For some, the message was displayed too late and for some too early.

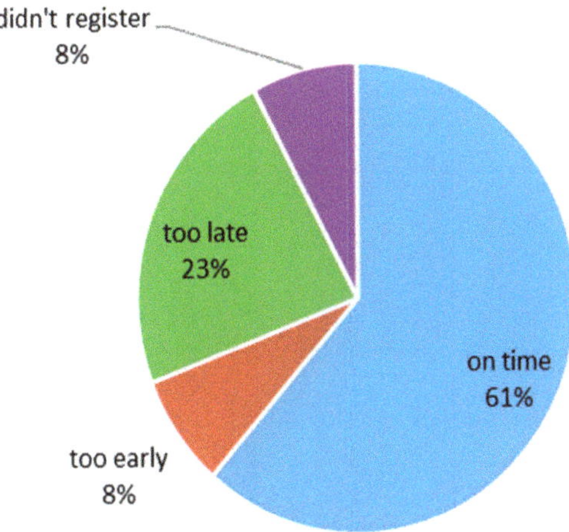

Figure 7. Information registration.

The post-ride answers are listed in Table 2. Unfortunately, the HMI distraction worsened compared to the pre-ride results in the static parameters. The most common value is no longer 4-disagree, but 2-agree. It clearly follows that HMI design must be focused on parameters like type, position, size, etc. The positive fact is the usefulness of the PTVC use-case is high (more than 50% strongly agree). Drivers also agree (with slightly lesser extent) that the information about approaching the crossing will increase safety.

Table 2. The results from post-ride questionnaires including specific questions on general and the use-case Public Transport Vehicle Crossing.

Post-Ride Questions Related to	Mean Value	Median Value	Mode	Variance	Standard Deviation
HMI distraction	3.23	3	4	1.03	1.01
Usefulness	4.31	5	5	0.90	0.95
Safety increase	3.85	4	4	0.97	0.99

Impact Assessment

For the public transport vehicle crossing, two main key performance indicators (KPI) were chosen to assess the difference in the driver's behavior with and without C-ITS warning. Two KPIs were selected according to the capabilities of the recording equipment and the data obtained during the evaluation of the PTVC. The first is the speed of the vehicle and the second is its acceleration. The data captured by the C-ITS evaluation logger was separated from the entire captured log according to whether it belonged to a significant zone for evaluating driver behavior.

The average driver's speed is shown in the box plot in Figure 8. This examines how the actual speed changed with the C-ITS warning about the approaching public transport vehicle. Comparing the two passages, we can state a speed reduction (approx. 4 km/h) during the passage with the C-ITS unit.

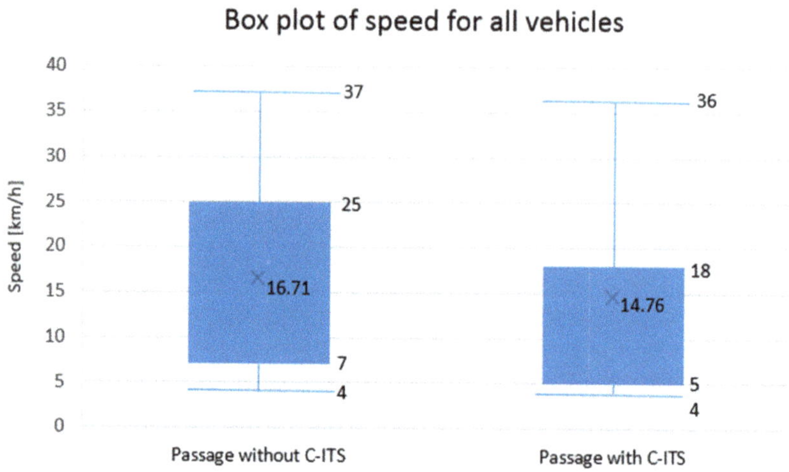

Box plot of speed for all vehicles

	Passage without C-ITS	Passage with C-ITS
Max	37	36
Upper quartile	25	18
Mean	16.71	14.76
Lower quartile	7	5
Min	4	4

Figure 8. Box plot of speed for all tested vehicles.

This trend could be also observed in Table 3 where we can see mean, maximum, and minimum speed in numbers. When driving with a C-ITS unit, drivers also had generally more uniform speed in common, which is visible on the smaller standard deviation (1.65 smaller with C-ITS) and the smaller range (by 1.17). Drivers tend to have lower average maximum speed (about 2.36 km/h) and lower average minimum speed (about 1.19 km/h) measured for all vehicles.

Table 3. Box plot of speed for all tested vehicles.

	Mean Speed [km/h]	Mean Max Speed [km/h]	Mean Min Speed [km/h]	Speed Range [km/h]	Standard Deviation [km/h]
Without C-ITS	16.71	30.26	5.70	33.43	9.36
With C-ITS	14.76	27.90	3.79	32.26	7.71

When performing a T-test of the statistical significance of the difference between the mean values with a 95% confidence interval, the resulting p-value is equal to 4.23×10^{-9} and we therefore consider the results to be statistically different. When performing Levene's test for the speed of both rides, the p-value of the test was 3.43×10^{-8}, so we can say that the probability of a difference caused by chance is minimal.

The second KPI selected for comparison was driver acceleration which is shown in Figure 9 and Table 4. Acceleration comparison does not show large differences with the use of a C-ITS unit. In both cases, the acceleration is centered around zero acceleration. The maximum and minimum acceleration values do not indicate significant braking or acceleration. When using a C-ITS unit, greater uniformity of driver acceleration can be seen on the range (1.35 m/s^2 smaller with C-ITS) and standard deviation (0.17 m/s^2 smaller with C-ITS), as well as lower average braking before without a C-ITS device. The sharp braking, which is taken when the deceleration of 5 m/s^2 is exceeded during the evaluation rides, was not exceeded in either of the two passes.

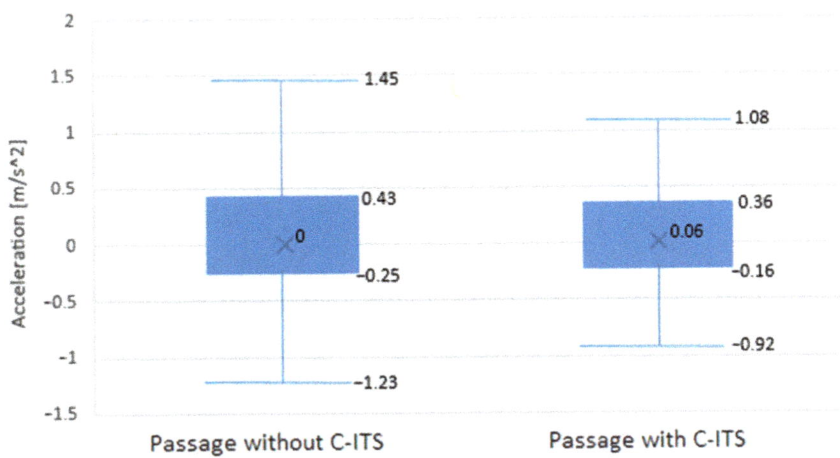

Figure 9. Box plot of acceleration for all tested vehicles.

Table 4. Box plot of acceleration for all tested vehicles.

	Mean Acceleration [m/s²]	Mean Max Acceleration [m/s²]	Mean Min Acceleration [m/s²]	Acceleration Range [m/s²]	Standard Deviation [m/s²]
Without C-ITS	0.000406	1.15	−1.73	4.65	0.69
With C-ITS	0.059209	1.17	−1.05	3.30	0.52

On performing a T-test of the statistical significance of the difference between the mean values with a 95% truth interval, the resulting *p*-value is equal to 0.13, and the mean value of the acceleration can therefore be caused by a random phenomenon. On the other hand, when applying Levene's test for the speed of both rides, the *p*-value of the test was 2.16×10^{-5}, so we can say that the probability of a difference caused by chance is very small.

Functional Evaluation

The results of the functional evaluation are divided into three parts according to the methodology.

1. Lessons learned:

The evaluation was carried out at a place where a tram comes out of dense forest. Because of that, it sometimes happened that the C-ITS message was not received in time. One solution could be a retransmission of RSU directly at the intersection.

There are also few recommendations filled out by the drivers:

- to inform the driver of the remaining time to the tram crossing,
- to inform on the tram speed.

2. Quality of service:

As mentioned in Lessons learned, the main issue was the late retrieval of information on crossing caused by insufficient event coverage. This issue negatively affects other important technical parameters such as latency and accuracy.

3. HMI:

For illustration, the warning about the public transport safety crossing is depicted in Figure 10 where the drivers are generally satisfied with the information from HMI while some drivers would welcome a larger pictogram of this use-case.

Figure 10. HMI of Public transport vehicle crossing use-case.

Figure 10 also shows the distance to the event, which is 22 m. At the time the vehicle is waiting for the tram to pass, there is the word "Now" instead of distance. Some drivers would simply omit this word.

The general comments regarding HMI also show that one of the problems that needs to be addressed in the future is the size and position of the HMI to inform the driver.

4. Discussion

The main objective of the C-ITS evaluation is to assess whether notifying the driver of the condition of an incident on the road in front of the driver will improve traffic safety, improve driving comfort, and readiness for the approaching event. A questionnaire survey was conducted to gain insight into the driver's views on C-ITS, the HMI, and the overall warning system.

This can be seen in the example of the PTVC use-case and its usefulness, and implementation together. Before the ride, the drivers were slightly inclined to believe that they would be distracted by the warning, which was favored by a few more drivers even after the ride. As the end user of C-ITS is a car driver, this should be seen as a good opportunity to obtain data and feedback from surveys on possible improvements and shortcomings of C-ITS and the impact on drivers.

The main result of the Impact Assessment analysis was a slight increase in caution when driving with C-ITS with an average reduction in driver speed, together with greater uniformity of measured speed and acceleration. It also turned out that the drivers braked less but also accelerated less using the C-ITS equipment. A big challenge is the effort to increase the accuracy of the measured data by filtering out external influences affecting the driver during the evaluation. As it is not possible to create the exact same driving conditions for all drivers, such as the impact of other traffic on drivers or the weather, it is necessary to take such factors into account in the analysis and reduce their effect on the overall conclusion by a larger number of evaluated passes. If an effort is made to balance the conditions for all evaluated drivers as much as possible, for example, by road closure or closed-circuit testing, there is a possibility that we expose drivers to an inauthentic environment in which they will not behave as they would in a real event.

Functional evaluation tells us important findings obtained during the evaluation, which significantly correlated with the opinion of drivers on the evaluated system. In this way, the late incoming messages and the insufficiently large pictograms depicting warnings were detected. One of the important outputs is the following implementation of important findings in C-ITS and its HMI from both functional evaluation and user acceptance. For the subsequent expansion and improvement of the methodology, the possibility of using more vehicles and thus multiple numbers of tested drivers for the same use-case in the same place and time are required. In this way, there is a better opportunity to capture a true view of the C-ITS system for wide publicity, which allows for better subsequent implementation of corrective measures in the HMI and the entire C-ITS system.

5. Conclusions

The main benefit of the article is the creation of evaluation methodology including specifics in the Czech Republic that are in line with the Evaluation and Assessment plan and FESTA handbook.

The methodology is divided into three parts: Evaluation preparation, Pilot Site Evaluation, and Evaluation and Assessment. Furthermore, Evaluation and Assessment are divided into three parts, User Acceptance, Impact assessment, and Functional Evaluation, Each part of the methodology is described and explained within the article and shown by an example of one specific use-case.

User acceptance showed the driver's positive view of the tested PTVS use-case, together with its usefulness, and implementation. The display of the HMI and the distraction of drivers following the displayed warning proved to be a significant factor in the evaluation of C-ITS, as well as a challenge for future studies. Impact assessment indicated a positive impact on the driver's behavior by reducing speed and slowing down more often when using C-ITS. To better understand and validate driver behavior on C-ITS systems, future studies should focus on streamlining the evaluation process for larger numbers of drivers.

Reference evaluation tests with specific responses to the event were conducted according to the proposed methodology and subsequently evaluated. As the results of the reference tests corresponded to the specified parameters, we consider the methodology to be valid.

Generally speaking, the evaluation should not be underestimated in terms of newly developed systems. If we focus on the C-ITS area we can see that the conclusions from the example are essential for the further development of C-ITS.

Author Contributions: Conceptualization, M.V.; Testing and methodology, M.V., M.Š. and M.M.; data processing, M.M.; project administration and management, Z.L. All authors have read and agreed to the published version of the manuscript.

Funding: This research was funded by European Union (CEF), within the C-ROADS project, grant number C-Roads_CZ_2015-CZ-TM-0188-M.

Institutional Review Board Statement: Not applicable.

Informed Consent Statement: Informed consent was obtained from all subjects involved in the study.

Data Availability Statement: Two types of data were used in this article: questionnaire survey data and logs recorded during the evaluation. The data from the questionnaire survey are covered by the GDPR Act and therefore it is not possible to publish them. The data that were used to evaluate the impact assessment are available from the authors on official request.

Acknowledgments: The published outputs are supported by voluntaries who joined the testing procedure and shared the personal data for the evaluation. The testing was also performed with the support of project partners (e.g., support of the back-office and control of infrastructure equipment).

Conflicts of Interest: The authors declare no conflict of interest.

Abbreviations

BO	Back Office
CAN	Controller Area Network
CEF	Common European Framework
C-ITS	Cooperative Intelligent Transport Systems
C-ITS SIM	Cooperative Intelligent Transport Systems Simulator
COOPERS	Co-operative Networks for Intelligent Road Safety
DPO	Public Transport in Ostrava
FESTA	Field opErational teSt supporT Action
FOT	Field Operational Tests
GDPR	General Data Protection Regulation
GPS	Global Positioning System
HMI	Human Machine Interface
KPI	Key Performance Indicator
OBD2	On-Board Diagnostics II
OBU	On-Board Unit
PTVC	Public Transport Vehicle Crossing
RSU	Road Side Unit
SCOOP	Pilot project for the deployment of C-ITS
WG	Work Group

References

1. Srotyr, M.; Zelinka, T.; Lokaj, Z. Pilot applications of cooperative systems. In Proceedings of the 2016 Smart Cities Symposium Prague (SCSP), Prague, Czech Republic, 26–27 May 2016.
2. Srotyr, M.; Zelinka, T.; Lokaj, Z. Hybrid communication solution for C-ITS and its evaluation. In Proceedings of the 2017 Smart Cities Symposium Prague (SCSP), Prague, Czech Republic, 25–26 May 2017.
3. C-Roads Platform. Harmonized C-ITS Specifications for Europe—Release 1.7. 2020. Available online: https://www.c-roads.eu/platform/about/news/News/entry/show/release-17-of-c-roads-harmonized-c-its-specifications.html (accessed on 28 August 2021).
4. FESTA Handbook Version 7, December 2018. Available online: https://www.connectedautomateddriving.eu/wp-content/uploads/2019/01/FESTA-Handbook-Version-7.pdf (accessed on 28 August 2021).
5. Crockford, G.; Netten, B.; Wadsworth, P. Establishing a common approach to evaluating the InterCor C-ITS pilot project. In Proceedings of the 2018 IEEE 87th Vehicular Technology Conference (VTC Spring), Porto, Portugal, 3–6 June 2018; pp. 1–2.
6. Innamaa, S.; Koskinen, S.; Kauvo, K. NordicWay Evaluation Outcome Report: Finland. 2017. Available online: https://uploads-ssl.webflow.com/5c487d8f7febe4125879c2d8/5c5c02add707426eba903956_NordicWay%20Evaluation%20Outcome%20Report%20M_13%20(secured).pdf (accessed on 28 August 2021).
7. Farah, H.; Koutsopoulos, H.N.; Saifuzzaman, M.; Kölbl, R.; Fuchs, S.; Bankosegger, D. Evaluation of the effect of cooperative infrastructure-to-vehicle systems on driver behavior. *Transp. Res. Part C Emerg. Technol.* **2012**, *21*, 42–56. [CrossRef]
8. Böhm, M.; Fuchs, S.; Pfliegl, R.; Kölbl, R. Driver behavior and user acceptance of cooperative systems based on infrastructure-to-vehicle communication. *Transp. Res. Rec.* **2009**, *2129*, 136–144. [CrossRef]
9. Malone, K.; Innamaa, S.; Hogema, J. Impact assessment of cooperative systems in the DRIVE C2X project. In Proceedings of the 22nd World Congress on Intelligent Transport Systems, Bordeaux, France, 1 January 2015; ERTICO-ITS Europe: Brussels, Belgium, 2015.
10. Esposito, M.C. Scoop@ idf: Implementation of cooperative systems for a road operator. *Transp. Res. Procedia* **2016**, *14*, 4582–4591. [CrossRef]
11. Agriesti, S.; Gandini, P.; Marchionni, G.; Paglino, V.; Ponti, M.; Studer, L. Evaluation approach for a combined implementation of day 1 C-ITS and truck platooning. In Proceedings of the 2018 IEEE 87th Vehicular Technology Conference (VTC Spring), Porto, Portugal, 3–6 June 2018; pp. 1–6.
12. Barnard, Y.; Innamaa, S.; Koskinen, S.; Gellerman, H.; Svanberg, E.; Chen, H. Methodology for field operational tests of automated vehicles. *Transp. Res. Procedia* **2016**, *14*, 2188–2196. [CrossRef]

13. Festag, A.; Le, L.; Goleva, M. Field operational tests for cooperative systems: A tussle between research, standardization and deployment. In Proceedings of the Eighth ACM International Workshop on Vehicular Inter-Networking, Las Vegas, USA, 23 September 2011; pp. 73–78.
14. Elhenawy, M.; Bond, A.; Rakotonirainy, A. C-ITS safety evaluation methodology based on cooperative awareness messages. In Proceedings of the 2018 21st International Conference on Intelligent Transportation Systems (ITSC), Maui, HI, USA, 4–7 November 2018; pp. 2471–2477.
15. Fank, J.; Knies, C.; Diermeyer, F.; Prasch, L.; Reinhardt, J.; Bengler, K. Factors for user acceptance of cooperative assistance systems. A two-step study assessing cooperative driving. In Proceedings of the 8th Tagung der Fahrerassistenz, München, Germany, 22–23 November 2017.
16. Kauer, M.; Franz, B.; Schreiber, M.; Bruder, R.; Geyer, S. User acceptance of cooperative maneuverbased driving—A summary of three studies. *Work* **2012**, *41* (Suppl. S1), 4258–4264. [CrossRef] [PubMed]
17. Liu, F.; Pueboobpaphan, R.; van Arem, B. Assessment of traffic impact on future cooperative driving systems: Challenges and considerations. In Proceedings of the 2009 International Conference on Ultra Modern Telecommunications & Workshops, St. Petersburg, Russia, 12–14 October 2009; pp. 1–5.
18. Jizba, T. Influence of HMI ergonomy on drivers in cooperative systems area. *Acta Polytech. CTU Proc.* **2017**, *12*, 42–49. [CrossRef]
19. Agudelo, A.F.; Bambague, D.F.; Collazos, C.A.; Luna-García, H.; Fardoun, H. Design guide for interfaces of automotive infotainment systems based on value sensitive design: A systematic review of the literature. In Proceedings of the VI Iberoamerican Conference of Computer Human Interaction (HCI 2020), Arequipa, Perú, 16–18 September 2020.
20. C-Roads Platform. Evaluation & Assessment Plan. 2019. Available online: https://www.c-roads.eu/fileadmin/user_upload/media/Dokumente/C-Roads_WG3_Evaluation_and_Assessment_Plan_version_June19_adopted_by_Countries_Final.pdf (accessed on 28 August 2021).
21. Lokaj, Z.; Srotyr, M.; Vanis, M.; Broz, J.; Mlada, M. C-ITS SIM as a tool for V2X communication and its validity assessment. In Proceedings of the Smart Cities Symposium Prague (SCSP), Prague, Czech Republic, 27–28 May 2021; IEEE Press: New York, NY, USA, 2021; ISBN 978-1-6654-3033-3.

Article

Comprehensive Analysis of Housing Estate Infrastructure in Relation to the Passability of Firefighting Equipment

Pavel Vrtal, Tomáš Kohout *, Jakub Nováček and Zdeněk Svatý

Department of Forensic Experts in Transportation, Faculty of Transportation Sciences, CTU in Prague, Konviktská 20, 110 00 Prague, Czech Republic; vrtal@fd.cvut.cz (P.V.); jakub.novacek@fd.cvut.cz (J.N.); svaty@fd.cvut.cz (Z.S.)

* Correspondence: kohout@fd.cvut.cz

Citation: Vrtal, P.; Kohout, T.; Nováček, J.; Svatý, Z. Comprehensive Analysis of Housing Estate Infrastructure in Relation to the Passability of Firefighting Equipment. *Appl. Sci.* **2021**, *11*, 9587. https://doi.org/10.3390/app11209587

Academic Editor: Leon Rothkrantz

Received: 9 September 2021
Accepted: 11 October 2021
Published: 14 October 2021

Abstract: The article focuses on the assessment and evaluation of the passability in densely populated parts of cities with multi-storey housing estates, in terms of the operation of the integrated rescue system (IRS) in the Czech Republic. The aim of the research is to minimize the arrival times to conduct the intervention as efficiently as possible. The presented problem is caused by unsystematic development of housing estates and the emergence of secondary problems in the form of inability to reach the place of intervention by the larger IRS vehicles. The vision presented in this document presents a systematic approach to improve the serviceability of individual blocks of flats. The main aim is to ensure passability, even for the largest equipment, such as fire engine ladders. Detailed mapping of the selected sites by drones, construction of their digital model, and subsequent virtual verification of the passability by specific vehicle models on identified access roads was performed. The results obtained by this procedure can then be implemented in the navigation of the fire safety forces and facilitate their arrival at the site of intervention. At the end, specific ways are presented in which the whole system can be modified to be able to intuitively change and choose individual access routes in real time, based on the current situation in the area.

Keywords: fire rescue service; smart cities; housing estates; swept paths; digital model

1. Introduction

The resilience of each city is determined by the level of risk management or the management of potential critical situations. In order to be effective in its function, the city itself must have the necessary comprehensive resilience to the potential threats, collapses, or functional limitations. One of the most important aspects of any city is the provision of transport infrastructure for the operation of the IRS. Specifically, this involves ensuring accessibility in densely populated areas of the city with high-rise prefabricated housing [1].

The history of housing estates in the Czech Republic varied and the architectural character of these localities has evolved considerably over the years. From an urban, architectural, and technical point of view, the largest problems are presented by the larger-scale housing estates. These were built mainly in the 1970s and 1980s and were almost exclusively constructed with prefabricated apartment buildings. An example is Prague, the capital of the Czech Republic. There are more than 600,000 flats in apartment buildings, in which almost 90% of the population of the capital city lives [2]. At the same time, the available statistics [3] show that almost half of Prague's population is concentrated in housing estates. There are currently 16 housing estates on the territory of the capital city with at least 10,000 inhabitants. Such large housing estates pose a number of traffic and transport challenges that are not currently thoroughly solved [1].

If we look at the housing estate as a system, it is possible to notice its gradual and continuous development, which in many aspects can be seen as a positive change (development according to the needs of residents), but also in many aspects as a negative one (unsystematic reconstruction of the area, and deviation from the originally intended transport and pedestrian links). The increasing level of motorization, and with it the ever-increasing demand for parking spaces [1,4], are not insignificant factors. For this reason, traffic at rest is accumulating in places that are not allocated for this purpose in the first place. Typical examples include inappropriate parking at crossroads, at the mouth of slip roads, or where parking is prohibited by traffic signs. Inappropriate parking reduces the efficiency of the transport links in the area and not only significantly reduces the traffic service but also worsens the safety and accessibility of the area.

The inappropriate construction layout of the area, together with the violation of road traffic rules, generates a significant problem, namely, the restriction of the passage of the IRS units during the intervention. Significant is the deterioration in time efficiency during passage through the area; thus, a delayed possibility of rapid intervention. These problems mostly affect the Fire Rescue Service (FRS), which has the largest vehicles within the IRS fleet. The length of fire tankers is almost 10 m, and in the case of a fire engine ladder truck, the vehicle is more than 10 m long [5,6]. Removing obstacles or finding an alternative detour route leads to significant time delays, which consequently results in increased risks to the person in danger or damage to the corresponding property.

The purpose of this article is to present an approach that can be used to solve the most important problems that limit the movement of the FRS through existing residential areas and then to point out the most important aspects that must be taken into account for a systematic evaluation. By applying once-off area-based data collection, it is possible to build a digital model of the area in which the quality of individual access roads can be assessed. On the basis of this analysis, it is possible to define a scheme of the key emergency routes along which the time losses are minimised and the passability of the FRS vehicles is ensured. Finding the most suitable access routes to each high-rise building allows the most effective and safe intervention. Additionally, effectivity can be increased by implementing the findings in the currently used software platforms by the FRS, e.g., dispatch and navigation [7]. A secondary objective is to provide examples of how to modify the whole clearance assessment approach using modern technologies in such a way that it can operate online in the future and intuitively assess risks in real time.

In the following chapters, current methods and procedures related to the passage of the FRS vehicles in urban areas and the methods aimed at mutual communication of the intervening vehicles are discussed. Furthermore, the methods used in the analysis of area in close proximity to the intervention target are discussed, followed by specific examples, justification, and the outputs of a pilot project. The next part is focused mainly on the visualization and creation of models related to the FRS ladders and their subsequent application in the search for the optimal solution within the area. The paper concludes with suitable methods to support procedure automatization in the course of the following years.

The following paragraphs describe the research method of the proposed project.

- Primary inputs: The motivation for the research was an impulse from the FRS, which has long struggled with the issue of the passability through housing estates by FRS vehicles. Due to the fact that this is a pressing and socially important issue, a working group was formed with experts from both the academic and practical sector. Among others, the aims of the research are the creation of a comprehensive methodological procedure for a complex analysis of housing estates in terms of the serviceability by FRS vehicles, to provide a definition of the appropriate procedures and methods for these analyses, and identification of the corresponding remediation measures.
- Dataset extension: In order to set up a suitable mechanism for the procedure of the research process, it was necessary to determine the methods that can be used to obtain a representative data set, which will then be subjected to detailed analyses. Closely related to this was the choice of appropriate measurement equipment and software to enable the collection of relevant data, their interpretation and analysis, operations with these data, and subsequent validation of the proposed solutions. At the same time, it is essential to choose methods and measuring instruments that can be potentially used in any location.

- Relations and model definition: In the next phase, the hypothesis for the specific outputs was defined. This involves the interpretation and analysis of the data obtained, defining the key parameters affecting the resilience of the housing estates accessibility in the research area.
- Outputs: The main objective of this research is to set up a suitable methodology for a comprehensive analysis of FRS vehicle passability in housing estates together with determination of the corresponding remedial measures. However, due to the complexity of the issue, it is also necessary to define the roles and steps for various involved stakeholders.

2. The Current Approach

The current approach to the problem of firefighting vehicles throughput, route planning, or effective arrival times focuses primarily on the routes between the fire station and the general area near the location of the fire. Several scientific publications have dealt with this topic, and several of them are summarised in the following paragraphs. Notable is a publication from the city of Changsha (China) [8,9] that is addressing the arrival times of firefighting units within the city. The premise in this study is that at least one fire tanker and one fire ladder truck should arrive at a given location within 240 s of a fire being detected. However, problems caused by congestion and a low density of fire stations in city centres make this assumption very difficult.

A similar approach was applied in the city of Nanjing (China) [10] in a study to verify the accessibility of different parts of the city in relation to the effective travel times of existing fire stations. The assessment of the suitability of fire station locations and their time of accessibility to the fire scene was analysed using the FC2SFCA method, which focuses on specifically defining the service boundary from each fire station and determining from which stations it is appropriate to lead the response to different parts of the city. The paper furthermore describes the comparison of the previously mentioned method with the older E2SFCA method and assesses the differences and precision of each procedure. The method shows that the coverage near fire stations or in the city peripheries is far more effective than coverage in the city centre, where high traffic density is a particular problem.

Publications from Pennsylvania State University (USA) [11] discuss the arrival times to the location of an incident together with a main focus on the impacts and economic losses associated with the time lost by the emergency services in reaching the fire location due to traffic congestion.

Other research related to integrated emergency services [12] is concerned with improving the efficiency of passage through light-controlled intersections, where emergency vehicles share their positional information with traffic lights. Research conducted in Saudi Arabia uses a mobile application that links the vehicle data with map data and the traffic light sensors. In the event that an IRS vehicle arrives at a traffic light, the vehicle's lane switch to the clear phase. In this way, traffic from other traffic directions does not accumulate at the crossroads and the traffic flow is smooth, without significant delays.

Research at Bangladesh University [13] focused on dynamic changes in the signal depending on the current position of the IRS vehicles. The proposed system works on a cloud centric Internet of Things (IoT) principle. It collects information about the location of vehicles and signals at traffic-light-controlled crossroads. The acquired information is evaluated, and the system subsequently changes the phases on the traffic lights to enable a priority pass-through of the IRS vehicles.

Another research study was trying to achieve similar results by solving crossing problems using a complex RFID-based system [14]. The system is designed to be able to detect both image and sound from the IRS vehicles and thus be able to change the order of phases at light-controlled intersections. This enables to the IRS vehicles an easier crossing of these incriminated points.

The research projects also aim to address the interaction between vehicles and their environment. A paper published by the University of Žilina (Slovakia) [15] presents cooperation between individual vehicles and sensor data, helping to increase the safety and improve the communication skills of individual road users. The use of vehicle-to-vehicle (V2V) and vehicle-to-everything (V2X) methods is intended to help warn drivers of approaching emergency services in time to create space for mutually safe crossing.

It could also be mentioned that there is a research focused on modelling buildings and their subsequent servicing by firefighting vehicles. Research from the National Central University of Taiwan [16] focuses on simulating fires on individual floors of high-rise buildings and their subsequent firefighting strategies. The proposed method involved the exploration of a 3D geometric network model (GNM) based on a building information model (BIM) to simulate firefighting. The operation of firefighting ladders during a response was also simulated in a 3D virtual environment. Finding the ideal location for the fire trucks to stop was supported by a methodology that can be used to determine the ideal location of the truck placement before the actual arrival of the engine in the fire area. At the same time, future research considers the use of real-time traffic sensors data to predict traffic conditions and adapt the optimal routing of the intervention line.

Similarly to the abovementioned strategy of deploying fire trucks around burning buildings, the research published by Belgian, Dutch, and Chinese academics [17] focuses primarily on water refilling of fire trucks in the process of fighting oil fires. The procedure uses the eM-plant simulation tool to model the firefighting process. In the context of the so-called chemical accident and to prevent its escalation, strategies are also considered to allocate fire trucks according to the distance from the hydrants and to divide fire trucks according to their number.

Improving the strategy of serving a given area is also discussed in a publication from the Netherlands [18]; it aims towards the optimisation of the allocation of firefighting vehicles to the location of an accident from the perspective of the emergency control room, based on historical dispatching problems where multiple vehicles from different locations are dispatched to the location of an incident and a decision has to be made as to which fire engine will be on site first. In this paper, a method to dispatch fire trucks is described in a way that strikes the right balance between rapid response to the actual fire and also good coverage of any simulated incidents.

After reviewing the currently available publications, it can be concluded that areas in close proximity to the actual intervention site are not being addressed in greater detail. However, these areas may have the highest loss times when the IRS vehicle is not able to access the actual location of the intervention. Currently, the intervention of the IRS forces is guided by the fastest possible route to the place of the critical accident, but without any knowledge of the actual situation in the target area. The innovation of the presented concept lies in the application and combination of knowledge from the Road Safety Audit (RSA), new methods of detail analysis of the area, especially by using aerial photogrammetry, and subsequent simulation in a virtual environment.

3. The New Approach

To further elaborate and verify the abovementioned conclusions, a pilot project in cooperation between the CTU in Prague, Faculty of Transportation Sciences, and the Prague 11-Chodov district was carried out. This project enabled the conversion of the theoretical intentions related to the solution of housing estates from the perspective of the passage of the FRS into real practice. This project has also enabled to define a methodological procedure that contains generalised guidelines to implement proposed solution in other localities. Figure 1 below shows the generalised steps of the procedure. The aim is to meet all the input requirements raised by the IRS units and local municipalities and to minimise any adverse effects on the residential needs. Marked boundaries around the steps, either solid or dashed, represents their nature. While the parameters marked with solid lines are determined and defined once during the evaluation. The dashed marked are iteratively adjusted over time in relation to the actual situation and development within the target area.

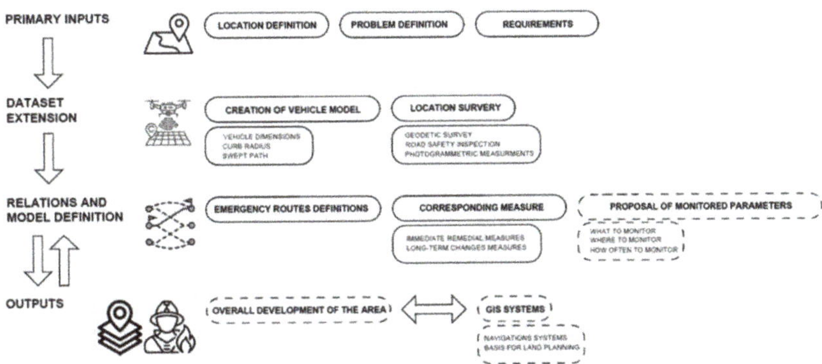

Figure 1. Diagram of the generalised procedure steps.

The target objective of the current project was to create a simplified offline digital model of the location, in which it would be possible to verify and evaluate the operation of the IRS during fire or other emergency situation. In the context of interaction between the FRS dispatching centre and the FRS commanders, this method can be used to prevent traffic complications caused in areas in close proximity to the target site and to avoid an increased arrival time due to inappropriately chosen routes for intervention.

3.1. Primary Inputs

In the beginning, it was necessary to assess all aspects related to the target area. Specifically, this involved assessment of the location and its relations within the town, identification of the key roads, and evaluation of their significance on the rest of the surrounding area. Additionally, it was necessary to define the appropriate crossings with the surrounding parent road network for the initial arrival of firefighting equipment. Linked to this is the necessity to assess the impact of traffic volumes on these roads, traffic flow mix, and junction capacity. These present an important factor that influences the initial choice of routes for intervention into the area. Furthermore, the assessment of the area from the point of view of the resident's demands was performed. The aim was to minimise the interference with the existing capacity of parking areas and at the same time with areas intended for leisure activities (parks and playgrounds). Lastly, it is necessary to consider the experience of the firefighters who serve the area, as they are aware of the most significant issues that limit the passage of their equipment.

3.2. Dataset Extension

In order to create a sufficiently detailed digital model of the area and to enable verification and simulation of the passability, low altitude aerial photogrammetry with a drone was used. The primary objective of the measurements was to obtain a digital three-dimensional spatial model with sufficient accuracy and detail. In contrast to classical measurement methods, the digital image correlation method was used, to facilitate the speed of data acquisition and processing [19]. The advantage of its application is that it does not require the explicit identification of individual points but uses image recognition algorithms to automatically identify the key features among the images, link them, and subsequently reconstruct the spatial information. Firstly, the Structure from Motion technique (SfM) was utilized. The SfM method uses a number of unordered images that depicts a static scene or an object from arbitrary viewpoints and attempts to recover camera parameters and a sparse point cloud that represents the 3D geometry of a scene [20]. The second stage consists of a further reconstruction of the geometric scene. This was carried by the Dense Multi-View 3D Reconstruction algorithm (DMVR), which was run on an already-aligned image set. The algorithm operates on a pixel base, thus enabling it to obtain a higher level of detail than the SfM [20]. The scale, transformation, and validation of the resulting model were performed with signalised points measured with the GNSS station. The resulting accuracy was 2.5 cm in position (X,Y) and 3 cm in height (Z). In this way, the resulting model was aligned with cadastral maps and it was possible to clearly define individual parcels intended for potential remediation. Figure 2 show a part of the assessed area in which the pilot project was carried out. Figure 2 depicts positions of captured images (over 3000 photographs were taken) together with the necessary overlap for successful reconstruction. Blue areas that are represented with the highest frequency represent the highest overlap of each image. The number of overlapping images then decreases to lower values in the extreme areas of the site. At the same time, the figure below also shows the statistics of the performed measurements.

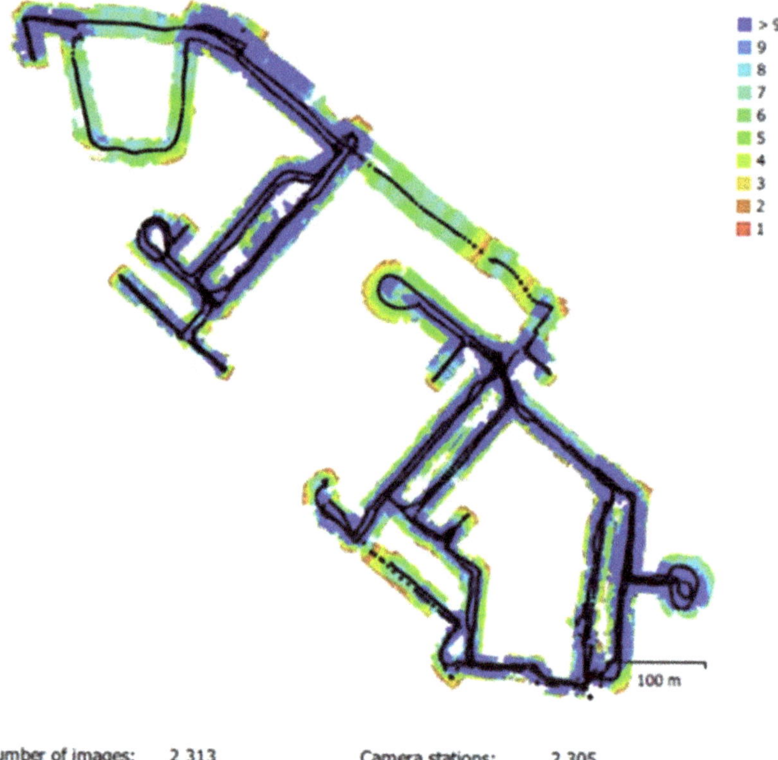

Number of images:	2 313	Camera stations:	2 305
Flying altitude:	22.1 m	Tie points:	5,297,607
Ground resolution:	6.72 mm/pix	Projections:	15,307,379
Coverage area:	0.107 km²	Reprojection error:	0.837 pix

Figure 2. Camera locations and image overlap with parameters.

The photogrammetric outputs are similar to laser scanning but with significantly lower time demands and corresponding costs, and only a slightly lower accuracy (RMS of 2.1 cm). For the purposes of this study, a textured triangular polygon mesh of the area was created together with a detailed orthophoto with a resolution of 2 cm/px [19,20]. Such detail is essential, especially in places where road width ratios reach very small values that border on the minimal required passage profile of the FRS vehicles.

The advantage of this approach is also the fact that the resulting model corresponds to the current situation at the time of measurement, e.g., shows parked vehicles, dustbin placements, fences, etc. Furthermore, it is complemented by the textures of the environment for effective presentation and orientation; therefore, it tells about the actuality and related deficiencies much more accurately than could be obtained from a panorama view, satellite imagery, or classical aerial maps.

In order to assess the passability, it is not necessary to work directly with a 3D model of the environment. The initial site analysis can be performed on a 2D basis (sufficiently detailed orthophoto is available). However, the use of a 3D model is necessary for evaluation of the clearance at locations where there are features that restrict the clearance profile of fire engines. These elements are, for example, unmaintained vegetation, road signs, streetlights, or bridge structures.

An illustration of the digital model is presented in Figures 3 and 4. Figure 3 shows the resulting spatial model, mesh, together with the location of each captured image. This presents a significant advantage due to the possibility to easily retrieve detailed photographs of the selected area during the subsequent analysis. Figure 4 shows the same model with textures.

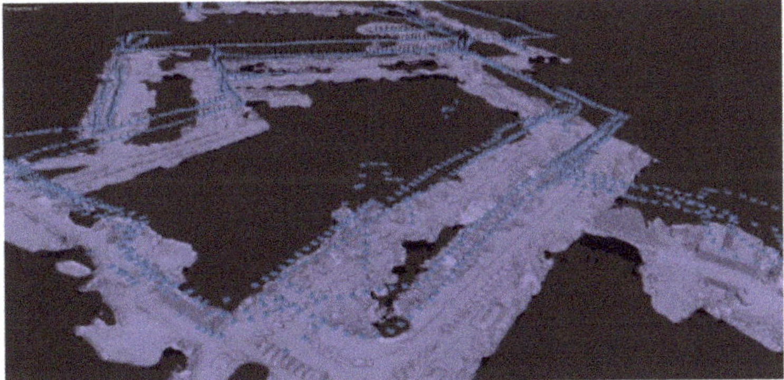

Figure 3. Digital model of the area with individual images.

Figure 4. Digital model of the area with textures.

Another way to expand the input of the evaluated area was through implementation of the RSA principles. The site inspection enables to detect potential critical locations and problems that might not be apparent at first look from the digital model or to somebody without traffic safety knowledge. Examples include high kerb edges at the entrances to emergency routes that could be used by the FRS, inadequate placement of road signs, or traffic-related safety issues.

Due to the specific geometric requirements of the FRS equipment, especially fire engine ladder trucks, which have the highest spatial demands, it is necessary to evaluate the passability of the vehicle through the area for the purposes of optimal intervention routes. Furthermore, the passability assessment enables verification of the efficiency of the selected routes. The passability can be easily verified with use of specialized swept path software. However, since almost every firefighting vehicle is unique in its parameters and differs from factory vehicles, the general vehicle data sheets cannot be used without adjustments. The same issue is with use of the predefined models designed for the verification of passability. A typical example of limited applicability of using already constructed models is the fact that each vehicle has different additional equipment, e.g., in the form of a pull-out ladder or a rescue basket. On the other hand, the basic precondition is that the high-rise firefighting equipment serving the area remains unchanged for several years (15–25 years). Therefore, the assessment or proposed changes can be related to the demands of specific fire vehicles that are serving the evaluated area.

As a result of incomplete technical documentation of the fire engine ladder truck used in the project (affected by the unique accessories mentioned above), it was necessary to create a model for subsequent simulations. A wide range of measuring methods can be effectively used to collect the data concerning the car body, its add-on components, and its dynamic characteristics, e.g., curb radius. The combination of the above-obtained parameters can then be suitably used in the creation of digital models, which can then be used to construct swept paths to verify the passability of the site under consideration. The chosen reference vehicle is shown in Figure 5.

Figure 5. Photo of a fire engine ladder truck [5,6].

3.3. Relations and Model Definition

The standardised vehicle models and freely available vehicle databases very often represent an idealized vehicle, which in its dimensions covers approximately 85% of the fleet of the same or similar type of vehicle or vehicles defined by national standards [21]. Common practice for passability assessments is to use the garbage truck model. However, due to significant differences in the dimensions and characteristics of the vehicle swept path during cornering, this leads to a situation where the vehicles that play the most crucial role in saving lives are omitted. Therefore, during the project, a simplified model of an FRS ladder truck was created and subsequently used to the assessment. Figure 6 shows the resulting simplified model defined in the AutoCad Vehicle Tracking software. For the purposes of model creation, it was necessary to determine not only the characteristic dimensions of the vehicle but also the actual turning radius, wheel angle, etc. The characteristic dimensions were determined with photogrammetric and geodetic measurements. Figure 7 shows the captured images of the ladder truck, reconstructed spatial model, and dimensions of the selected parts of the truck (length, height, ladder dimensions, etc.).

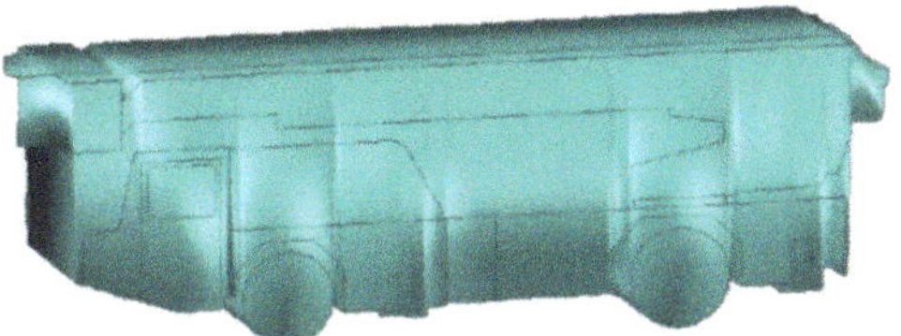

Figure 6. Final digital model of a fire engine ladder truck created in AutoCad Vehicle Tracking.

Figure 7. Photogrammetric measurements of the characteristic dimensions of the FRS ladder truck in Photomodeler Scanner.

Figure 8 shows the differences between the garbage truck model and FRS ladder truck.

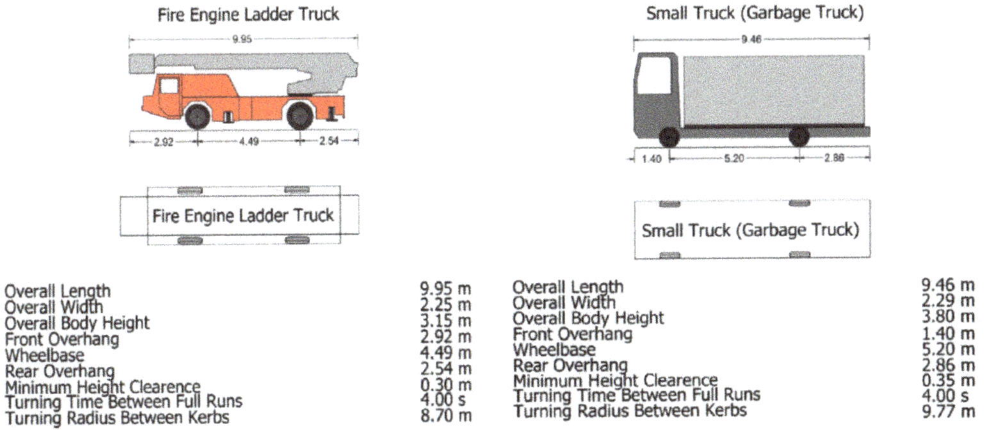

Figure 8. Technical parameters of the fire engine ladder truck and the garbage truck.

In order to obtain a real swept path of the reference FRS vehicle, a research experiment was conducted with an FRS vehicle (Figure 9). In this experiment, the full turn of the FRS vehicle at maximum wheel travel was monitored by drone. The speed of the vehicle was determined based on the FRS experience at approx. 5–10 km/h. The footage was subsequently rectified to enable direct identification of wheels trajectory and the vehicle contours. Thus, it was possible to determine the actual turning radius and wheel angle.

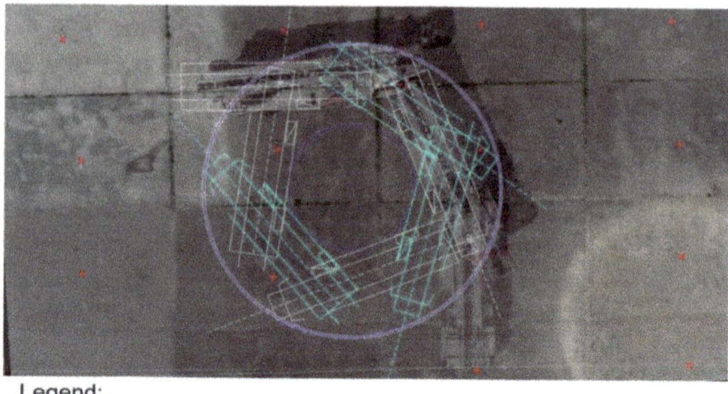

Figure 9. An example of a calibration experiment of the swept path for vehicle model design.

Ackermann's steering geometry was subsequently used for verification of the model validity, specifically by calculating the steering angle. The principle is based on a fixed axle where only the wheels rotate and is shown in Figure 10 [22].

$$\alpha_i = tan^{-1}\left(\frac{L}{R - \frac{T}{2}}\right) \tag{1}$$

$$\alpha_0 = tan^{-1}\left(\frac{L}{R + \frac{T}{2}}\right) \tag{2}$$

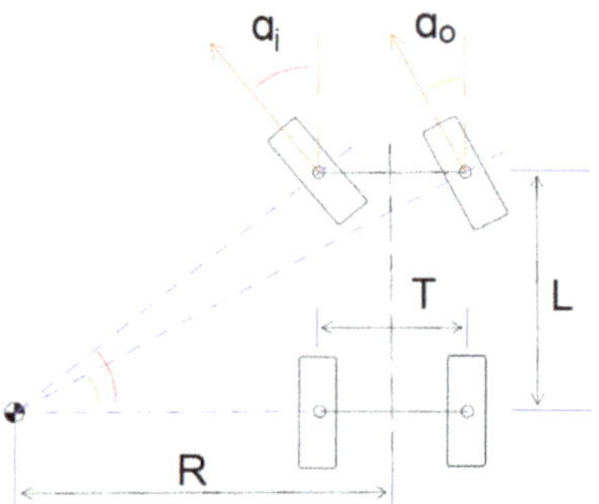

Figure 10. Ackermann's steering geometry [22].

By substituting the observed turning radius into the equation, a value for the wheel angle was obtained and compared with the experimentally measured value for the wheel angle. This verified that the determined vehicle turning radius corresponded to the real vehicle.

The resulting simplified model showed that vehicles that are very similar to firefighting vehicles in terms of their parameters (length, width, and height) actually have significantly different trajectories. Figure 11 shows a comparison of the swept paths of a municipal waste vehicle (blue) and a fire engine ladder truck (red). This difference in their manoeuvrability is closely related to the different turning radius or steering angle of the wheels. The different behaviour can be seen especially in the initial phase of the turning manoeuvre, due to the greater body overhang of the fire engine ladder truck compared to a conventional truck.

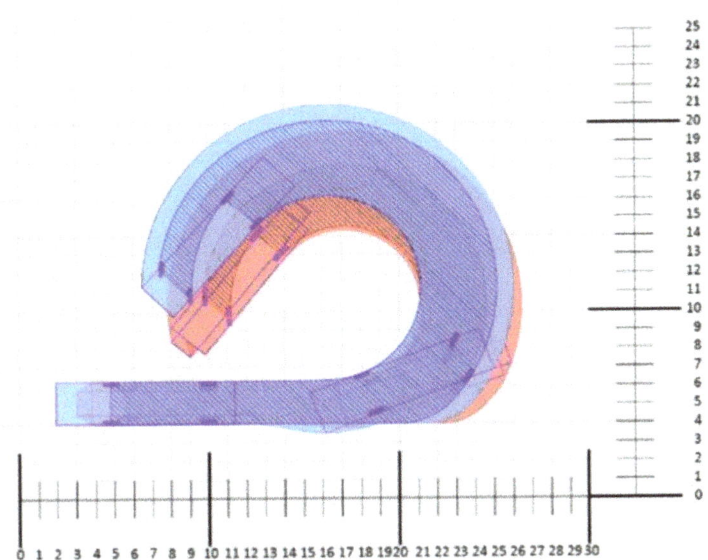

Figure 11. Swept paths of the fire engine ladder truck and garbage truck.

By using digital models of firefighting equipment and a digital model of the area, it was possible to verify the passage through the selected area in the current layout; this enabled to define the critical locations where potential bottlenecks may occur and which require subsequent structural modifications. These basic outputs became the basis for subsequent comprehensive research outputs. The most frequent problems affecting the passability are high kerb lines, which cannot be driven over into the vicinity of buildings; incorrectly implemented parking places, which interfere with the crossroad areas; inappropriately placed vertical traffic signs, which prevent parking of vehicles with the overlap of the bow onto the pavement; or inappropriately placed solid obstacles, e.g., concrete posts, railings, etc. The main objective is to eliminate the identified problems as soon as possible. Priority is given to problems that can be solved easily, at a low cost, or with low corresponding administration. However, the general idea is that the measures must be performed systematically and in coordination between all involved partners.

Most prefabricated buildings in the area have several possibilities from which a rescue operation can be performed, but not all roads are suitable for the passage of larger firefighting equipment, e.g., spatial or load limitations. The most efficient solution logically seems to be the use of roads that were designed to enable firefighting operations at the time the area development. However, there are cases when during time the purpose of these roads changed, some of the connections disappeared, or parts of them are currently used for different purposes than the one for which they were originally established. Examples include the creation of parking places, flower gardens, partial or total redesign, etc. Despite this fact, these roads can still serve as a basis for the area serviceability.

Based on the available infrastructure in the evaluated area, several options were proposed to ensure the accessibility of every individual building in the project. At the same time, it was possible to determine the theoretical time delays of the individual intervention routes depending on the travel time to each location. Figure 12 illustrates the serviceability options for the area and the design of the main or alternative emergency route. Furthermore, this solution identifies the locations where it is possible to perform only a one-way passthrough or where it is possible to turn around or pass another vehicle. The digital model also provides information about the vertical distance of the road from the face of the building. This distance significantly influences the overall reach of the fire ladders during an intervention and thus the course of the actions. It is essential to always find an area in front of the building where the distance is as small as possible, but at the same time, the area is wide enough to allow the support legs of the vehicle to be extended as far as possible to achieve the required stability and is able to withstand the load of the vehicle.

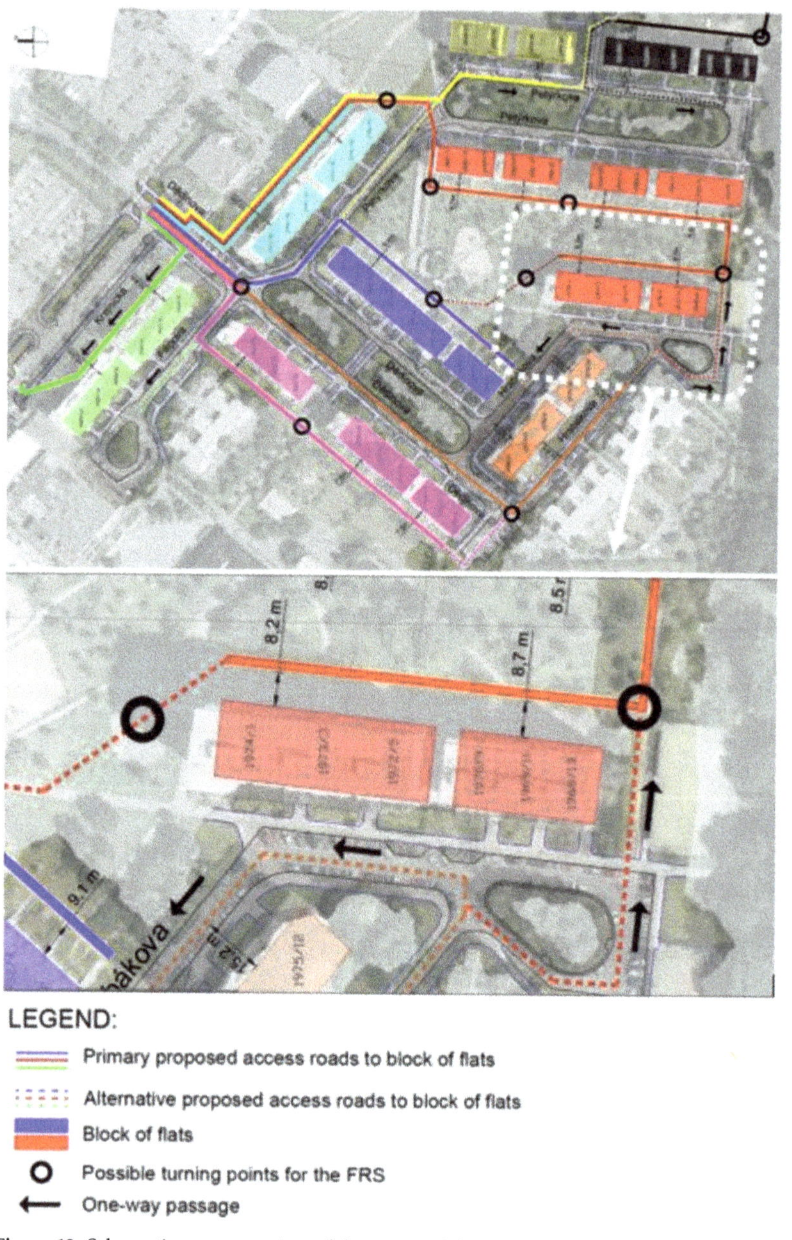

Figure 12. Schematic representation of the serviceability proposal for the evaluated area together with details of the selected area.

Each entrance road to a particular block of flats has different loss times and different levels of complications depending on the occupancy of the roads by parked vehicles. For this reason, it can be presumed that time losses of an intervention vary during the day and directly correspond with the actual situation. During working hours, when most of the residents are away from their homes, the time-losses are lower in comparison to the evening, when parking capacity is considerably saturated and vehicles are parked even in places not set up for this purpose. For this reason, it is desirable to supply and accordingly modify the proposed emergency routes with online information, e.g., in the form of camera footage or by using data from traffic sensors. The advantage of the proposed approach is that the potential bottlenecks or critical locations can be determined in advance. Thus, the monitoring is designed with consideration of the specifics of the evaluated area. This leads to the possibility to select the most appropriate access route to individual buildings in a significantly more effective way than through a simple empirical experience.

3.4. The Validation of Proposal and Project Outputs

A mathematical description of the presented problem of FRS passability through housing estates can be determined by two basic premises. The first premise is the attempt to minimize the time of arrival of the FRS to the location of the fire, similarly as in case of optimization of serviceability of larger areas. The second premise works with the hypothesis of maximisation of the sum of probabilities for effective FRS intervention, such as maximisation of the probability of collision-free passage through the area in such a way that there are no elements limiting passage along the route, such as the spatial layout of the roads. Minimalisation is the probability that potential traffic excess in the form of a bottleneck will occur on the determined access route. The third factor is the probability of effective intervention directly at the place affected by the fire in terms of the technological possibilities of FRS vehicles. This is in close relation to the specific parameters of the target building/object. Parameters such as the maximum possible height available by means of a fire truck ladder or the minimum perpendicular distance from the extinguished object must be considered here.

However, there is a significant difference between the optimization of serviceability of larger area, e.g., part of the city, and the blocks with high-rise buildings, where only few potential routes are possible, special infrastructure is already in place, and only few access points exist. Therefore, it is very difficult to perform specific mathematical verification of the proposed routes, as each FRS intervention is very unique and different complications may occur. It is only possible to obtain a theoretical verification, but this lacks many variables and the result is thus highly biased. The general verification procedure of the proposed changes is related to the experimental determination of the transit speed of each route. For this reason, it is necessary to perform control measurements directly in real traffic with specific cars. In the authors' opinion, a general mathematical prescription for verifying the passage of FRS vehicles through a specific location cannot be made explicit using a clearly defined mathematical apparatus. Each such passage is unique and is always influenced by a number of variable parameters.

The research found that the FRS may not be the only end-entities benefiting from the research. Other involved parties can benefit from the research and use this for the city's development and to increase its safety. One of the key users is the dispatch of the IRS, which can implement the information in the navigation of the emergency services based on the actual place of the emergency. At the same time, it enables to inform the responding units in advance about the current traffic situation, possible risks, or most effective access routes. With the help of the online digital model of the area, it is possible to provide reliable information about the state of the site, both in terms of traffic and the possibilities of the vehicle placement. At the same time, it is possible to provide information where it is possible to pass another vehicle or the most suitable places for staging fire trucks so that the intervention can be carried out from the closest possible distance.

The identified service routes are also directly applicable to the fire truck crews. The Czech FRS is currently using a map-based application for localisation and best utilisation of the hydrants in the area of intervention. The services routes are prepared and implemented using GIS systems. This approach enables to add this information within the application to improve the information available to the fire truck crew. Furthermore, this approach also enables to have an up-to-date overview of the individual service routes in the area in case the monitoring system is active. A similar approach is reported in Portuguese research, which also implements a decision-making system in GIS applications for the crew of the FRS in order to reach the fire location as quickly as possible and thus reduce the consequences [23].

Last but not least, the GIS data can be used to inform the municipality administration about critical areas and suitable passage routes for the IRS. The benefit is a safe and systematic development of the area in the future, as it is known which roads are necessary or where the problems arise; also, it enables to mitigate potential disruption of links in the service of the IRS in cases of road closures.

4. Discussion

While the methodological procedures applied in the previous chapter can be modified to achieve a direct response to the actual situation in the locality, the practical implementation is limited in several ways. The changes outlined in this chapter summarise how to effectively supplement the digital model of the area and improve its serviceability based on online information.

The most important part of the project, which currently contains the most data, is the digital model of the location. To be able to talk about a real virtual copy, which is similar in its characteristics to the vision of the Twin City model [1], it is necessary to add elements and systems capable of real-time monitoring of the city's traffic processes thanks to sensors directly connected to this virtual model. This includes, for example, data from sensors used to calculate the current occupancy of parking places [1]. If it is clear that the capacity of the parking places is not fully utilised, there will most likely be no inappropriate parking of vehicles at crossroads and no significant disruption to the flow or throughput of traffic through the site. In the event that the parking capacity is fully occupied, inadequate parking can be expected even at locations that are crucial for passability of the IRS (crossroads, fire roads, etc.). In these cases, the use of camera monitoring and subsequent systematic control seems to be an ideal solution [24]. When monitoring a lot of similar locations, it is possible to use induction loops or detectors working on a similar principle as a support tool that automatically informs the control centre of a potential issue.

Another approach is the potential utilisation of drones, which can be effectively used for detailed and periodical analysis of the assessed locations in a very short time. Verification of critical places or areas could be done similarly to the small parcel delivery system introduced by Amazon in recent years. Figure 13 illustrates a docking station mounted on a lamppost to serve as both a take-off point, charging station, and a connection point for the dispatch centre controlling the area. The drones provide an ideal means to scan a selected area in a short time frame and send the acquired data to a central control station before the actual firefighting operation takes place. In this way, it would also be possible to optimise in advance the route option during the intervention, thus avoiding any significant time delays [8]. However, the possibility of using drones in built-up areas is limited. Currently, such a data collection option cannot be implemented, as legislation and data protection do not allow the use of drones over built-up areas [25,26].

Figure 13. Drone docking stations on lamp posts [27].

Another way to obtain the necessary data for the assessment and analysis of selected locations can be through lidar or ultrasonic detectors. These can be mounted on garbage trucks that visit the sites several times a week. The collection of these data could be used to update the input data and to update the area model itself.

5. Conclusions

The effort to systematically evaluate the target areas of intervention in densely populated parts of cities with multi-storey housing estates has very specific goals—serviceability by the IRS, reduction of the arrival times of the IRS, and to enable the passage of large FRS vehicles. The research presented methods for obtaining the necessary data to evaluate the current condition of the locality and to design remediation measures. Primarily, these were the application of various photogrammetric and geodetic measurements to obtain outputs that were subsequently used for modelling in specialised software. This outlined a suitable procedure for mapping the locality and for obtaining the specific parameters of the FRS vehicles. This contribution presented a novel methodological approach that contains generalised guidelines for their practical implementation. The methodology enables to create a digital model of the location, in which it would be possible to verify and evaluate the operation of the IRS during fire or another emergency situation. Furthermore, the approach was validated through a pilot study in cooperation with municipal administration and the FRS. The pilot project has made it possible to consider all the specifics of the selected location and proposed a comprehensive solution in the form of a methodological procedure that can be used repeatedly for the analysis and subsequent design of measures for high-rise housing estates.

The new methodology approach of the area management also helps during the future development of the area, as it is possible to determine which roads or areas in the locality are crucial to ensure the serviceability by the IRS or which presents the highest risk or time delays. For this reason, it is also possible to plan a systematic remediation of the weak spots. The vision of the conceptual design was primarily to create a digital 3D model in which individual variants of the firefighting equipment's arrival to the intervention site can be simulated, verified, or tested, depending on the actual situation in the area. By integrating it into the current navigation systems used by the FRS, it is possible to optimise the arrival time and minimize the loss of life or the level of damage to properties. At the same time, the methodological approach presented by the research can be further developed by extension of the spectrum of the measured data or by implementation of other approaches to collect these data; for example, using sensors, camera surveillance systems, or detectors, as discussed in the previous chapter.

The safety and ideal serviceability of existing urban areas is a continual process, which needs to be periodically revised and improved due to the area development. It is essential to apply new procedures and use modern technological possibilities to facilitate the work of the emergency services to improve the resilience of the urban area in relation to the potential loss of life and property.

In this paper, we define approaches and methods for the comprehensive analysis of housing estates in terms of FRS vehicle passability. The paper presents a detailed outline of the scientific approach of each step of the proposed solution itself. We are aware that the problem addressed does not only affect housing estates in the Czech Republic but is a problem occurring worldwide. From a historical point of view, this problem is reasonably expected; for example, in all post-communist countries of Central and Eastern Europe, where in the past similar housing estates were constructed from uniform prefabricated materials. As a result of this conclusion, we believe that the presented approaches and methods, while performed only in the Czech Republic, and thus with consideration to its specifics, are applicable with only slight adjustments globally. This statement is also supported by the set of technological resources and software available to academics and professionals globally. This research and its outputs provide a clearly defined and unified set of methods applicable to the analysis of any housing estate across the world's countries.

Author Contributions: Conceptualization, P.V., T.K. and J.N.; methodology, P.V., T.K. and J.N.; software, P.V.; validation, P.V., T.K. and J.N.; formal analysis, P.V., T.K. and J.N.; investigation, T.K.; resources, P.V., T.K. and J.N.; data curation, J.N.; writing—original draft preparation, P.V. and T.K.; writing—review and editing, P.V., T.K. and J.N.; visualization, P.V. and Z.S.; supervision, P.V., T.K. and J.N.; project administration, P.V., T.K., J.N. and Z.S. All authors have read and agreed to the published version of the manuscript.

Funding: This research received no external funding.

Institutional Review Board Statement: Not applicable.

Informed Consent Statement: Not applicable.

Data Availability Statement: Not applicable.

Conflicts of Interest: The authors declare no conflict of interest.

References

1. Svítek, M.; Michal, P.; Votruba, Z.; Přibyl, O. *Cities of the Future*, 1st ed.; NADATUR: Prague, Czech Republic, 2018; ISBN 978-80-7270-058-5.
2. Czech Statistical Office. Housewives. 2011. Available online: https://www.czso.cz/documents/10180/20567427/10413513k6.pdf/95ecab47-9596-472b-8132-f75b023bcc42?version=1.0 (accessed on 15 July 2021).
3. Czech Statistical Office. Settlement Structure and Housing Type. 2011. Available online: https://www.czso.cz/documents/10180/20567443/42413513a2.pdf/c1bb081b-740e-405f-bd56-2a556d865123?version=1.0 (accessed on 15 July 2021).
4. Technical Road Administration. Transport Yearbook Prague 2020. 2020. Available online: https://www.tsk-praha.cz/static/udi-rocenka-2020-cz.pdf (accessed on 16 July 2021).
5. Kratochvíl, M.; Kratochvíl, V. *Technical Equipment for Fire Protection*, 1st ed.; Sdružení Požárního a Bezpečnostního Inženýrství: Prague, Czech Republic, 2009; ISBN 9788073850647.
6. Jendřišak, J.; Svoboda, P. *Fire Trucks in Czech Republic*, 1st ed.; FIJEPO: Prague, Czech Republic, 2005; ISBN 80-902705-4-9.
7. Systemflorian.cz. Available online: https://systemflorian.cz/ (accessed on 5 November 2020).
8. Liu, D.; Xu, Z.; Yan, L.; Wang, F. Applying real-time travel times to estimate fire service coverage rate for high-rise buildings. *Appl. Sci.* **2020**, *10*, 6632. [CrossRef]

9. Liu, D.; Xu, Z.; Yan, L.; Fan, C. Dynamic estimation system for fire station service areas based on travel time data. *Fire Saf. J.* **2020**, *118*, 103238. Available online: https://www.sciencedirect.com/science/article/pii/S0379711220307566?via%3Dihub (accessed on 5 November 2020). [CrossRef]
10. Mao, K.; Chen, Y.; Wu, G.; Huang, J.; Yang, W.; Xia, Z. Measuring spatial accessibility of urban fire services using historical fire incidents in Nanjing, China. *ISPRS Int. J. Geo-Inf.* **2020**, *9*, 585. [CrossRef]
11. Daniel, B.; Louis-Philippe, B. Traffic congestion, transportation policies, and the performance of first responders, Pennsylvania, USA. *J. Environ. Econ. Manag.* **2020**, *103*, 102339. [CrossRef]
12. Almuraykhi, K.M.; Akhlaq, M. STLS: Smart Traffic Lights System for Emergency Response Vehicles. In Proceedings of the ICCIS 2019 International Conference on Computer and Information Sciences, Sakaka, Saudi Arabia, 3–4 April 2019; ISBN 978-1-5386-8125-1. [CrossRef]
13. Sangamesh, S.B.; Sanjay, D.H.; Meghana, S.; Thippeswamy, M.N. Advanced traffic signal control system for emergency vehicles, bangalore. *IJRTE Int. J. Recent Technol. Eng.* **2019**, *8*, 1242. Available online: https://www.ijrte.org/wp-content/uploads/papers/v8i3/C4323098219.pdf (accessed on 5 November 2020).
14. Raman, A.; Kaushik, S.; Rao, K.R.; Moharir, M. A Hybrid Framework for Expediting Emergency Vehicle Movement on Indian Roads. In Proceedings of the 2nd International Conference on Innovative Mechanisms for Industry Applications (ICIMIA), Bangalore, India, 5–7 March 2020; pp. 459–464. Available online: https://ieeexplore.ieee.org/document/9074933/ (accessed on 5 November 2020).
15. Jantošová, A.; Dolnák, I.; Dado, M. Cooperative and emergency services feasible in future vehicular networks, Žilina. In Proceedings of the ICECCE 2020 International Conference on Electrical, Communication, and Computer Engineering, Istanbul, Turkey, 12–13 June 2020; ISBN 978-1-7281-7116-6. Available online: https://ieeexplore.ieee.org/document/9179485 (accessed on 5 November 2020).
16. Chen, L.C.; Wu, C.H.; Shen, T.S.; Chou, C.C. The application of geometric network models and building information models in geospatial environments for fire-fighting simulations, Taiwan. *Comput. Environ. Urban Syst.* **2014**, *45*, 1–12. [CrossRef]
17. Zhou, J.; Tu, C.; Reniers, G. Simulation analysis of fire truck scheduling strategies for fighting oil fires. *J. Loss Prev. Process Ind.* **2020**, *67*, 104205. Available online: https://www.sciencedirect.com/science/article/pii/S0950423020304927 (accessed on 5 November 2020). [CrossRef]
18. Usanov, D.; van de Ven, P.M.; van der Mei, R.D. Dispatching fire trucks under stochastic driving times, Netherlands. *Comput. Oper. Res.* **2020**, *114*, 104829. [CrossRef]
19. Pavelka, K.; Faltýnová, M.; Švec, Z.; Dušánek, P. *Mobile Laser Scanning*, 1st ed.; CTU in Prague: Prague, Czech Republic, 2014; ISBN 978-80-01-05261-7.
20. AGISOFT LLC. Agisoft Metashape User Manual: Professional Edition. Version 1.7.0. 2020. Available online: https://www.agisoft.com/pdf/metashape-pro_1_7_en.pdf (accessed on 15 July 2021).
21. Transport Research Centre. *Swept Paths for Verifying the Passability of Road Elements*, 1st ed.; Ministerstvo Dopravy České Republiky: Prague, Czech Republic, 2004; ISBN 80-86502-14-7.
22. REIF. *Automotive Handbook*, 9th ed.; Konrad a Karl-Heinz DIETSCHE, Revised and Extended; Robert Bosch: Karlsruhe, Germany, 2014; ISBN 978-1-119-03294-6.
23. Lourenço, M.; Oliveira, L.B.; Oliveira, J.P.; Mora, A.; Oliveira, H.; Santos, R. An Integrated Decision Support System for Improving Wildfire Suppression Management. *ISPRS Int. J. Geo-Inf.* **2021**, *10*, 497. Available online: https://www.mdpi.com/2220-9964/10/8/497/htm (accessed on 5 November 2020). [CrossRef]
24. Hussein, T.M.; Erol-Kantarci, M.; Rehmani, M.H. (Eds.) *Transportation and Power Grid in Smart Cities*, 1st ed.; Wiley: Hoboken, NJ, USA, 2019; ISBN 978-1-119-36008-7.
25. Unmanned Aircraft—Civil Aviation Authority. Civil Aviation Authority—Safe and Sighted [Online]. Copyright © 2021 All Rights Reserved. Available online: https://www.caa.cz/en/flight-operations/unmanned-aircraft/ (accessed on 30 August 2021).
26. Cureton, P. *Drone Futures, UAS in Landscape and Urban Design*, 1st ed.; Routledge: Oxfordshire, UK, 2020; ISBN 9780815380511.
27. Multi-Use UAV Docking Station Systems and Methods. U.S. Patent 9,387,928 B1, 12 July 2016.

MDPI
St. Alban-Anlage 66
4052 Basel
Switzerland
Tel. +41 61 683 77 34
Fax +41 61 302 89 18
www.mdpi.com

Applied Sciences Editorial Office
E-mail: applsci@mdpi.com
www.mdpi.com/journal/applsci

www.ingramcontent.com/pod-product-compliance
Lightning Source LLC
LaVergne TN
LVHW070702170726
843515LV00004B/465

* 9 7 8 3 0 3 6 5 4 1 4 7 1 *